Staring into the Sun

By Nick Brunacini

About the Author

Nick Brunacini joined the Phoenix (Ariz.) Fire Department in 1980 as a firefighter. He retired as a Shift Commander in 2009.Today he teaches the Blue Card Hazard Zone Program with his brother and father. He also writes for B Shifter, the Hazard-Zone Quarterly.

Staring into the Sun

Copyright 2012 by Nick Brunacini &
Across the Street Productions Inc.

ISBN

Staring into the Sun is published by
Published by Nick Brunacini &
Across the Street Productions Inc.
3119 W Orangewood Ave.
Phoenix, AZ 85051-7448
www.bluecardcommand.com

Printed and bound in the United States of America.

Cover Art by Jason Anthony

This book is dedicated to the City of Phoenix, where I got to spend my 29-year firefighting career. Thank you for building me the most wonderful playground.

Chapter 1
Lead Feet & Lies

Every driver dreams of driving a fire truck. Imagine winning a lottery where the prize was getting to navigate a gleaming red fire engine through city traffic. The experience would almost certainly overwhelm the lucky recipient. The first time many firefighters get the opportunity to drive Big Red, they develop a severe case of "Jake leg."* When driving fire apparatus we are allowed to break traffic laws because we're responding to emergencies. This is expected of us. We are not lollygaggers. We are highly trained professionals who tread where angels fear. We must respond Code 3 or society will collapse. Firefighters are young and beautiful, possess cat-like reflexes and the strength of 10 ordinary men. This is because most of you are chock to the rim with testosterone (even the girls have some of the man hormone). The headiest cocktail on God's green earth is equal parts testosterone mixed with adrenaline. It's a very easy drink to overdose on. We must be very careful when the thrill juice floods our brain to not let it overwhelm our judgment and actions because the results can be disastrous.

*Jake leg is derived from Jake brake. A Jake brake is an engine brake used on some diesel motors. When the driver of a vehicle equipped with a Jake brake takes their foot off the gas it shuts off the supply of fuel to the engine—turning the motor into a compression brake. When you hear a semi truck barreling down the freeway slow down for a curve and the engine makes a loud rumbling noise, this is the Jake brake working. A Jake brake has a tendency to make the vehicle jerk because of complex properties of physics that I don't understand. New drivers unfamiliar with operating fire apparatus often times are very herky-jerky with the gas. Hence the reference Jake leg.

My brother John remains the only person I've ever known who accumulated enough points to have his driver's license suspended before being old enough to qualify for a learner's permit. One evening, the police pulled him over for speeding. He was 14 years old, and it was around 3 a.m. The cops wrote him a ticket and then allowed him to drive home. My brother did severe damage to several vehicles prior to qualifying for his learner's permit. Most of these were single-vehicle, unreported affairs. Looking back on it, I estimate he and his buddies did tens of thousands of dollars in damage while they were learning the nuances of vehicle control.

One day, my father attempted to drive his old work truck. It was a 1951 Chevrolet five-window, half-ton, step-side pickup. This creampuff weighed as much as two brand new Chevy pickups, was six different colors and started with a screwdriver. Dear old dad climbed in, inserted his straight-blade screwdriver into the ignition switch, stomped down on the starter and nothing happened. He figured it was a bad battery, so he got out of his workhorse and popped the hood. To his dismay, the battery was broken in half. On further inspection, he found the front axle was bent backward almost a foot. The questioning began with me. I had an airtight alibi, so next up was John. Initially John blamed it on car thieves. My father had some snappy reply like, "Let me get this straight: Criminals stole my truck, proceeded to bust it up and then were courteous enough to tow it back to my house?"

Seeing that his current lie was going nowhere fast, my brother quickly tried to shift the blame to my 12-year-old sister. "Candi did it because she knew eventually I would get blamed for it. She blames me for everything." This was a bad tactic. My sister was still recovering from an incident the previous week when my brother threw her off the roof of our house, breaking her arm. I could tell my father was losing his patience. He was momentarily lost in some private thought as he stared at my sister and her swollen, white tongue. (She had a plaster cast on her arm that she continually licked. The family felt the fall caused her to think she was the family cat, Cookie, who had disappeared a year earlier. None of the therapists were ever able to properly diagnose her condition, but that's another story.) It is a miracle my father didn't stuff my brother into a dry-cleaning bag. The truck incident ended with a stern lecture and 30 days of manual labor.

My brother quit driving my mom and dad's vehicles for a month or so, but that didn't stop him from serving as a passenger in his scofflaw buddies' parents' vehicles. John got his final ticket during a late-night drive in the park behind our house. The park was recently irrigated and had anywhere from 6 inches to a foot of water covering its 10-acre surface. My brother's good friend Rob had "borrowed" his father's new Blazer. It was a beautiful four-wheel-drive vehicle that had been lifted and had oversized, off-road tires and wheels. It also had lots of aftermarket engine components and produced far too much horsepower for hormonally challenged boys. They called their park driving "Rat Patrol" for the wildly popular TV show about testosterone role models—fighting soldiers. Many of you will remember the show was based on Army men driving jeeps equipped with machine guns, which they used with great patriotic delight, sneak-attacking the Nazis during the WWII African campaign. The boys were having a blast, taking turns behind the wheel making rooster tails in the stagnant park irrigation as they did donuts. The fun ended when they swamped the motor in the middle of the park. The cops were waiting for them. This ticket put my brother over eight points, which would have caused the suspension of his license—if he had had a license in the first place.

That was his final ride of the summer. A couple weeks later, with his latest deed well behind him and the promise of a new school year ahead, my brother signed up for driver's education. He learned many fascinating things, stuff like traffic laws and how to signal a turn. It was all very elementary for someone who knew how to fishtail a car around a corner and how to hit a berm in the middle of a park full of water to get the most air under your tires. Class ended, and John applied for and actually received his learner's permit. The only explanation I can offer is the state didn't do background checks on people who weren't old enough to possess a permit or a license. This all changed when he made it through six months with his learner's permit, trouble and ticket free, and applied for his driver's license. The lady at the DMV told him that he passed his driving test with flying colors, and was such a nice young man, but there was a problem with three or four outstanding tickets and he would have to go to juvenile court to appear before a judge before the state would issue him a driver's license.

4

John is very fortunate that I've always looked several decades older than my actual age. He had been very successful hiding his tickets from our parents, and he pleaded with me to represent him in court as his father. One thing led to another, so I put on my dad's shoes and one of his jackets and had my brother drive me to the juvenile court and detention facility. I looked very stoic and didn't have to show ID or speak very much, so the meeting was short and sweet. My brother threw himself on the mercy of the court. After paying a $75 fine, John was legal to drive.

You may be asking yourself what this has to do with driving fire apparatus. I submit to you that it has everything to do with fire apparatus. Many of the firefighters I know have similar histories. Three years after my brother lied to a judge and I impersonated a person who I could never be, we were both driving fire apparatus in the city of Phoenix. We were legally breaking every traffic law that had ever been written. The engineers who taught us how to drive fire trucks were experts. They could do things with a double-clutch, manual four-speed, 16-ton fire truck that would awe any NASCAR driver worth his salt. They had also accumulated their share of traffic "mishaps" during their formative years. We all grew up in a culture where the best drivers were also the fastest drivers. The old guys tempered their speed with equal parts experience and luck. The young guys relied almost entirely on reflex and luck. It is an accurate observation that for the most part we were all out of control.

I struggle with the concept of a "standard response."

Early in my career, standard response meant we went balls to the wall on every single call. It didn't matter if it was a working fire, man down, kid in a pool or a cat stuck in a tree. When the engineer turned on the siren, you could expect the truck to be on two wheels at some point during the ride to the scene. After a while it dawned on me that it was probably unnecessary to respond with so much gusto to calls we categorized as "bullshit." Remember those pre-enlightened days when we used that vulgar term to describe any incident that didn't require us to risk our lives to save the customer's life when we showed up? If the call's thrill level didn't exceed the out-of-control ride on the way there, it was pretty much filed away in the bullshit category. I began to wonder why we drove to bullshit calls like we had been launched out of a steam catapult from the deck of an aircraft carrier. I never heard an explanation that made sense: "You never know what you are really responding to until you get to the scene," "This is the way we have always done it," "Shut up and hang on" and "Blah, blah, blah." Back in those days, the enlightened and wise few who actually had the nerve to challenge the ancient ways made the argument that if we were responding to a real deal, life-or-death incident, it was very important to actually arrive on scene. These heretics believed our response shouldn't cause more harm than the call we were responding to. These forward-thinkers were not well received; the mob thrives on tradition, hard-ons and false deities—not logic.

The engineers I worked with were very good at

driving and operating their fire trucks. If they weren't, they would have been in an accident every single shift. They never outperformed their driving skills, but they routinely outperformed every other motorist's driving skills. This is one of the problems with driving a large red vehicle with lots of blinking lights and loud sirens in a highly aggressive manner at a high rate of speed. It has a tendency to scare other drivers, oftentimes causing them to take inappropriate action. It was commonplace to hear engineers say they could always control their vehicle (my brother used to say this a lot when he was 15 years old). I never doubted them (but I knew John was overestimating his abilities at the time). The trouble is they can't control what the other drivers are doing and how they react when our engineer breaks half a dozen traffic laws in the span of 300 feet while traveling at speeds in excess of 60 mph. After an accident where we tore the back end off an old lady's Buick, I remember the engineer saying he could have missed the old woman if she would have just stopped like she should have. Driving in most urban areas is dangerous enough when you follow all the traffic laws.

Testosterone, adrenaline and culture still held that good engineers could smoke their way into other companies' areas and be first due. Nothing had changed since the days when the horses pulled the apparatus. These three essential ingredients (testosterone, adrenaline and culture) must be combined with maturity and logic. You ever taste unsweetened chocolate? It has all the flavor of dirt until you add the sugar, then it's transformed into pure magic. In the absence of sufficient amounts of logic and maturity, our hormones and tradition cause us to do some real stupid shit. Firefighters should never drive a water tanker (or any other vehicle) in a manner where it can roll over and kill them. When was the last time shaving 20 seconds off your arrival time made the difference between someone living and dying? Was it even this year?

I was a battalion chief at a fire station that housed two engine companies. An old engineer drove one of the trucks. He was slow and steady. He had worked in this area for 30 years and knew where everything was. The other engineer was younger and very adept at driving harder (i.e. faster) than his older counterpart. The older engineer loved to shine his truck. The younger engineer was very mechanical and used to screw with the old guy by throwing nuts and bolts under his rig. The old man would spend hours trying to figure out which part of his truck they had fallen from. These two rigs would routinely respond together to structure fires. The younger engineer always got out of the station first and drove faster toward the incident. He never understood how the old engineer always ended up parked in a better spot and his crew invariably got first water onto the fire. The old guy knew where he was going and how he was going to get there before he ever left the station. The young guy was fueled by testosterone and smoke on the horizon. We always followed the old guy.

I read a number the other day that was outrageous (I found it at http://www.jpl.nasa.gov/releases/2000/eviewsmonrovia.html). From the early 1980s to 1995,

U.S. emergency-response vehicles were involved in more than 156,000 traffic accidents, resulting in 6,550 deaths. I don't advocate removing the lights and sirens from our apparatus or taking our sweet ass time when we respond on calls. But we can find a happy balance between getting to the incident scene in a timely manner and leaving a swathe of destruction in our wake. Seven years after I served as my brother's legal guardian and attorney, he was driving fire trucks full time as an engineer, and I was riding on the tailboard. Our sister had finally recovered from her fall off the roof and had quit urinating on the wheels of my parents' cars. John survived his early driving career with a couple minor accidents that led to the revision of several of our fire department's driving policies. John was very fortunate that his driving mishaps only resulted in the city buying some unfortunate motorist a new car. He quickly figured out there wasn't a happy future in placing speed over control and went on to become an excellent driver.

For most of us, this essay (and any other driving discussion) is a lot of theory and conjecture. I know firefighters who haven't been so lucky. They've been involved in accidents where some of the participants didn't survive. No one escapes these events the same. The one lesson we can take away from these stupid tragedies is to manage our response in a manner that eliminates their possibility altogether. The only way to do that is by slowing down and always stopping when we don't have the right of way. We all say that no building is worth a firefighter's life. If that's true, then our response isn't worth killing anyone over.

Chapter 2
Born Again

A few days ago I read that a fire department is going to stock EMS gear for pets on all of its rigs. Imagine carrying four sets of equipment: pediatric, adult, doggy and kitty cat. Five years ago, it was common to hear firefighters bemoan that we should stop diluting our "craft & trade" by taking on the other types of services. Now some departments have added emergency veterinary service to their repertoire. Certainly a sign of the end days.

I've been involved in incidents where we've rescued both humans and animals. The human rescues are merely a sidebar in the story. If someone has photos or video of a saved family pet, the human element may not even be reported. Let's face it, we love our pets. Outside certain family members and a few friends, I'm more attached to a small mutt my family took in two years ago than I am to the other 6 billion people I share this planet with.

When you combine a pet rescue with a miracle birth, you'll surely make headlines and be the toast of the town. And if the saved pet is an exotic one from a faraway land, well you just couldn't ask for much more, and the story will take on a life of its own. I was lucky enough to bear witness to one such occurrence. Like most historic and life-changing events that take place across the plane of human existence, this one happened in an unassuming and modest area of the city.

It was just after 5 a.m. and two engines, a ladder company and a battalion chief had been dispatched to a house fire. The officer on the first-in engine reported heavy smoke showing from the rear of a medium-sized house. They were pulling an attack line for search, rescue and fire control and assumed command of the event. The crew was in the process of flaking their line in the front yard when they encountered a hysterical woman. "Get my babies out! Don't let my babies die!" she begged.

The incident commander (IC) reported over the tactical radio channel that they had reports of children trapped inside the house. Just like everyone responding to the call, I woke all the way up and turned up the volume on my radio. The first-due ladder company was the second unit to arrive on scene. This unit's captain was (and continues to be) a very serious individual. He's a no-nonsense kind of guy. He reported his arrival to the scene and told the IC over the radio that he and his crew were headed to the roof to perform vertical ventilation. The IC told him no, they had the fire knocked down and there was very little heat in the house. They needed positive-pressure ventilation and assistance completing the search.

Captain Serious parroted this order back to the IC and went to work. I arrived on the scene in my battalion chief (BC) wagon a minute or so behind the ladder company and transferred command. At this point, we had seven highly motivated firefighters inside a medium-size house searching for dying children. We balanced the incident out to a first alarm in the event there were people in the smoke-filled house. A man stood with the woman in the front yard. He was trying to comfort her. Every time she dropped to her knees, he would pick her up.

She would stand and sob for a few seconds, then drop again. He would pick her up once more. In the background, light white smoke pushed out of the house behind the screaming noise of the 10 hp motor on the vent fan. Crews on the interior radioed that they had achieved fire control but hadn't found anyone inside the house. The second arriving engine company reported that they were staged on a hydrant half a block to the south. I ordered them to make access through the backyard to see if there were any kids who had sought refuge to the rear of the house and then to assist with completing the interior search. We were 3 or 4 minutes into the call when the sun was starting to come up, and I could hear a news helicopter overhead.

The third engine had just staged north of the scene when the cell phone in the BC rig rang. It was our department's public information officer (PIO). The media had heard the radio reports that children were trapped in the burning house, and they wanted the juicy details. My partner gave him the nickel version, took down his number and told him we would call him back in a few minutes with fresh information. Interior crews reported that they just finished the third sweep of the entire house and turned up no victims. The crew we assigned to check the backyard had just opened a side gate, directly across from the command post, and was shepherding out all form of animals. There was a Billy goat, sheep, dogs and cages with rabbits and birds. Cats were attacking the cages, trying to get at the birdie and bunny breakfast treats. The front yard was quickly turning into Noah's Ark.

We reported the fire was under control, declared an all clear on the house and cancelled the balance of the first alarm (no victims, no need). I was watching the festivities in the front yard when Captain Serious knocked on my window. He said the fire started in a mattress that had filled the interior with white smoke. There was very little fire damage, but the place had been smoked up pretty good. The only life form they found inside the house was a furry rodent in a cage that the owner referred to as a sugar glider. Captain Serious thought it looked like an attractive rat. The animal was tits up when they found it, but a firefighter went above and beyond the call of duty and gave it mouth-to-mouth resuscitation. This brought the furry little critter back to the here and now. Captain Serious and I looked over to see the crying woman in the front yard clutching her "baby" and showering it with kisses. Before leaving, the Captain added, "One other thing, the rat had a couple of rat babies."

After we radioed that we hadn't found any dead people in the house, the incident turned into your run-of-the-mill house fire, and media interest dropped way off. I called our PIO to let him know one of our guys had actually revived the owner's unusual pet. "No people victims, the only thing the search turned up was the owner's sugar glider."

"What the hell is a sugar glider?"

"I'm told that it's a furry, rat-like animal from Australia."

"OK, what's news worthy about that?"

"When they found it, it was dead and one of our guys gave it CPR and brought it back to life."

"Well, that is interesting."

"That's not the half of it. After they brought it back to life it gave birth to a couple of sugar-glider babies."

"Liar."

"I heard it straight from Captain Serious on Ladder 24."

"Let me talk to him."

I waved Captain Serious over and told him what I told our PIO. He confirmed the basic details of the story.

"When we found the cage, I only saw the one big, hairy, dead rat in it. After my guy blew in its mouth a couple of times, it came around. Later on when I saw the cage there were a couple of babies suckling the mama rat." Captain Serious handed the phone back to me and disappeared in the front yard zoo. I went back to talking to the PIO.

"Unbelievable isn't it?"

"I have a bad feeling about this one, but it's just too good not to share."

One of the 6 a.m. newscasts had a 30-second blip about our house fire. Video shot from the news helicopter showed the petting zoo in the front yard—the main focus of the story. This was before our PIO put out a media alert describing the biblical event in a humble, everyman and folksy style. Twelve hours later, the sugar glider incident was the lead story on all of the local news channels. The following morning it made the front page of the newspaper. A day later the morning disk jockeys at the most popular country music station in the world wrote a song about the blessed event. The national wire services picked up the story. People living in remote parts of the planet were now privy to our little house fire.

The next day I was back at work and first thing in the morning I received a frantic phone call from Captain Serious. The media had learned his crew was responsible for saving momma rat and her rat babies. He told me he had to take his phone off the hook yesterday after the tenth media inquiry. The two of us formulated a plan where he and his crew could talk to the media one time and be done with it. Our PIO set up a live video shoot for the midday news cast. They would do it in front of the fire station that housed Captain Serious and his famous sugar-glider champions. When I drove by the station at 11:45 it looked like the O.J. trial. We couldn't find a place to park so we went back to our quarters and watched it on TV. Mr. PIO started things off and then introduced the ladder crew. I had the unnerving feeling that all of the members of the now-famous ladder crew were glaring at me through the TV. They confirmed they had in fact administered CPR to mama sugar and saved her life, but none of them had actually seen her deliver the two live sugar pups. The media didn't care; they had a miracle on their hands. This story was bigger than a vision of Christ appearing in a flour tortilla.

Things took a different turn that evening. A reporter for one of the TV news channels did something completely out of character for a news reporter—she actually investigated the story. She tracked down the homeowner and interviewed her on camera. The homeowner (sugar glider's master) was very grateful and full of praise for the firefighters that saved her baby. She went on to describe what a sugar glider is along with a bunch of details generally reserved for the Discovery Channel. When asked about the miracle birth the woman let the sugar glider out of the bag. She told the world that this was a big misunderstanding.

Mama sugar glider had the babies the day before the fire, and they had spent the night in another cage. After the firemen saved the mama, she (the homeowner) had put her back into the same cage with her babies. She could understand why the firemen mistakenly assumed the wondrous birth. The reporter was very good-natured about it and ended her story with a giggle. Some of the other news channels did not want to believe her. Many of them seemed angry that someone would let "facts" interfere with such an amazing and joyous story. Things were starting to heat up when another fire-related story knocked the sugar-glider event out of the news. A serial arsonist had struck again. This was an ongoing news event, containing elements of eco-terrorism, fire, the FBI, ATF, arson-sniffing dogs (always a big part of the story) and local firefighters. The arsonist had already torched four or five high-end houses under construction near the edges of mountain preserves. One story said this misguided individual felt he was protecting our open spaces. Several months later, after he was captured, it turned out that his house bordered the same mountain preserve. Another selfish, criminal moron.

During the course of sugar-glider week, the Phoenix Fire Department went on more than 3,000 calls. We delivered babies, treated and transported ill diabetics, put out structure fires, rescued trapped hikers on mountains, saved heart attack victims, along with all of the other routine stuff we do during an average span of seven days, just like thousands of other fire departments. The sugar-glider story overshadowed all of it until the continuing acts of an infamous and anonymous psychopath knocked it out of the spotlight. One firefighter had the compassion to blow some air into someone's dead pet. It didn't take a minute or cost a dime. The pet's owner only cared about her "baby" when we pulled up to her burning house. Captain Serious and his crew were able to give her the only thing she wanted during that frantic time in her life. The owner got what she wanted, and we got a million dollars worth of good will. What else is there?

Chapter 3
The Spud of Sound

A mother brings her young son by the local firehouse so the tyke can see the spectacular red fire engines and meet the firefighters. One of the firefighters is showing the little boy a rig when the kid tells him, "I want to be a fireman when I grow up." The firefighter replies, "Kid, it looks like you have a decision to make because you can't do both."

I have an associate who was briefly engaged to a girl whose mother was a psychiatrist. The mother was never fond of my friend. She used to tell her daughter that firefighters were borderline pyromaniacs who suffered from the Peter Pan complex. Their relationship hit rock bottom when the mother unexpectedly showed up one day and caught her other daughter giving her future son-in-law a bath. (It was a love that was never meant to be.) My friend was a fine firefighter and a great American. He has the intuitive ability to quickly exploit any situation to his benefit. This is a quality one looks for in a workforce that responds to life and death emergencies. You want workers who can quickly make sense out of chaos, develop a sane and lucid plan, then take quick, definitive action. One of the beautiful things about our work is that when we win, the customer wins.

The Phoenix Fire Department has hired many nontraditional firefighters in the recent past. These new hires abandoned good careers in favor of starting over. These professions included lawyers, stockbrokers, bankers, jet fighter pilots, accountants, concert pianists and cops. It has become commonplace to see 40-year-old recruits going through our training academy. The reason is simple: We have the greatest job on the planet. In my hometown, around 3,000 people show up every few years to compete for a hundred or so firefighter positions. Our Personnel Section claims it's more difficult to gain employment as a Phoenix firefighter than it is to get accepted in to University of Arizona Medical School. Our job is so good sometimes we forget it's a job. In most cases this is a good thing, but one of the pitfalls we face is the lure of inappropriate antics. We've all seen or heard about things getting out of hand and the antics blowing up in our faces. Looking back on some of these things, one can only wonder how we could have been so stupid. Factoring in a group of firefighters will generally cube the effect of a stupid antic gone wrong. A group contains much more stupidity potential than a single person.

The first time I ever saw a potato cannon was behind Fire Station 11. We had finished up with the dinner ritual and were out back playing horseshoes. It was the dead of winter, so the ambient air temperature hovered in the mid-fifties. To keep away the cold, most of us were clad in bright yellow brush jackets that we had been issued that summer. Station 11 is located north of Sky Harbor International Airport and east of downtown Phoenix. Our closest brush interface was 10 first due areas away, so we didn't go to many brush fires. We found the next best use for this fine piece of protective gear—campfire duster. Our group was a finely tuned, well-oiled machine, ready at a moment's notice to rush out into an uncertain world and go eye-to-eye with death and destruction. In the meantime, we were throwing horseshoes and embellishing stories around a fire pit. Many of the men were enjoying cigars. A new paramedic was waxing his $40,000 pickup truck. I told you, we have the greatest jobs anywhere.

One of the guys assigned to the ambulance walked over with a black piece of PVC pipe. It was 2" in diameter, around 4' long and open on one end. The other end had a 6" cleanout attached to it. The end cap of the cleanout had been drilled out, and an igniter from a gas barbecue had been installed in the hole. To load the potato cannon, a potato is shoved into the open end of the pipe and tamped down with a broom handle. Once the potato is properly loaded, the end cap is unscrewed and a 3-second blast of hair spray is directed into the cleanout section of the device. (I'm told Aqua Net is the preferred propellant.) After the end cap is screwed back into place, the apparatus is shaken to achieve the proper fuel-air mixture. When the cannon is fired, it sounds like a mortar going off, and the potato is quickly hurled farther than the eye can see. We didn't have a care in the world while we launched potatoes that evening; it was loads of juvenile fun. I wasn't concerned until the next shift, when I saw that one of these devices had the power to leave something much more significant than a welt. Someone taped a bull's eye to the side of a 40-gallon, galvanized steel trashcan. A potato cannon, fired from 15 feet away, crushed the can in half. This wasn't a toy one would give to a child. It became troublesome when the members of the station began to build their own potato guns in hopes of besting the next guy's. A couple weeks later, with no high-velocity potato mishaps (yet), the other station captain and I decided we were all pushing our luck. The B Shift crew had built an arsenal of cannons that could launch a potato nearly half a mile; we had to stop the madness. We quietly put a moratorium on potato cannons and went back to more mundane pursuits.

I'm sure some of you are shaking you heads in disbelief—the professional guardians of a community engaging in dangerous and sophomoric antics? Guilty as charged, but I think it's important to note that none of us ever fired a potato in anger or with bad intentions. It wasn't a very wise pastime, but hiring a group of people who willfully run into burning buildings comes with some management challenges. Our little firehouse family was fortunate that we stopped our illicit hobby before something dire happened. It's always better when we wise up and cease the silly antics on our own, as opposed to pressing forward, blowing up the apparatus bay and having to explain why a firefighter requires medical attention. It is generally painful when the fun and games go public.

It had been more than a decade since I saw my last potato cannon. I figured it died the same death as the mullet hairstyle, just kind of faded away. I was transported back to that time in my career when I heard an unsubstantiated rumor a couple of months ago. In the interest of decorum, let's just say the incident allegedly occurred somewhere in the Southwestern United States of America, and we can leave it at that. This is the story as I heard it: A group of young firefighters was experimenting with potato cannons. One of the more inquisitive members of the tribe had an idea to get more oomph out of his launcher. After he packed a specially selected spud down the barrel, he filled the combustion chamber with 100 percent oxygen before giving it a 3-second blast of hair spray. He screwed the cap into place and prepared himself for taking the shot heard around the neighborhood.

In hindsight it was fortunate that the crew took some precautions. They knew by adding pure oxygen to the potato-launching formula it would increase the force of the blast to the highest levels. According to the legend, some type of blind was used to shield the cannon master from his weapon. A small hole in the shield allowed him access to the igniter. The remainder of the crew took refuge behind solid objects. The stage was set, the weapon primed. The crew was about to make history. They were firing the cannon from the sanctity of the apparatus bay. The rigs were parked on the front apron; the front doors were closed. The rear bay doors were open and provided a shooting alley into the large parking lot, where potatoes met a grizzly death when they kissed a block wall after traveling at speeds in excess of 100 mph. Our hero depressed the firing mechanism, giving spark to the supercharged fuel mixture. If I had to guess, I'd say the potato came apart around the time it hit the speed of sound. Something certainly broke the sound barrier because the dozens of fluorescent lights and the window glass in the bay doors all exploded into tens of thousands of tiny shards of glass right after the doomed cannon blew into bits. When the glass settled, the only injuries were ringing in the ears and a small laceration received by our misguided potato master. Like all really good legends, this one has its share of unanswered questions. I don't know how the crew explained the hospital visit required to close up the small head wound. All the broken glass also presented a sizable problem. Did anyone get shifts off? I honestly can't answer that. None of this matters anyway.

Like any event that turns out badly, we can all take a lesson from this. We have fabulous careers. Three quarters of those who do it, do it for free. No one can ruin our tasty and delicious working conditions any faster or more completely than we can. The siren song of silly antics is strong, but for our own good we must resist. I always cringe when my bosses end long paragraphs with, "Where was the captain?" It's even worse when they replace the word "captain" with "chief."

Self-destruction is never pretty. Like most stupid things, there isn't a problem the first 100 times we do something idiotic. In almost all cases the problem bites us in the ass on the last time we do it. (For most reasonably intelligent people, when an activity leads to harm we cease doing it, hence, the last time.)

This group of firefighters wasn't the first to fire vegetables out of homemade mortars, and they probably won't be the last. Education is the key. By sharing lurid tales of mishaps we can all learn. We should never undertake any activity that has the potential of causing structural or biological damage.

24

Chapter 4
The Lawnmower Man

Most firefighters join the service with illusions of doing hand-to-hand combat with the products of combustion, saving nurseries full of cooing babies and being canonized by members of the opposite sex. However, we must examine the two realities about how we spend our days.

1) Whenever a structure fire breaks out in our village, we will be the ones who respond and fight the fire. You light 'em, we fight 'em.

2) These fires represent the small end of our service-delivery scale. At the end of a regular shift, today's firefighter will be wearing more vomit, blood and misplaced body fluids on their uniform than they will smoke and soot.

When I began my career, I was told the ratio of runs for the year was about 75 percent medical and 25 percent fire. The paramedic who taught our EMT training kept referring to the "golden hour," saying if we stabilized and transported the patient to a trauma center within an hour of their injury, their chances of survival rose dramatically. Our group graduated from the training academy ready to keep the city from burning down and prepared to chase death away from the sick and injured. No one told us that we would spend lots of time tending to the mentally infirm. These people's ailments were not measured in increments of golden hours. My initial experience with the mentally different happened during my first day as a new firefighter, assigned to an engine company in the southern area of the city.

I remember the excitement of my first shift. I had brand new uniforms—the latest fashion statement in turnouts and toiletries—and bedding, complete with a white crinkle bedspread from JC Penny (it was regulation). I introduced myself to the members of the station. My captain told me where I would be riding on the 1978 Mack pumper and what my job responsibilities were for structure fires, serious car accidents, shootings, stabbings and routine EMS calls. This area of town was south of the Salt River (actually a dry riverbed peppered with wrecking yards and concrete production facilities) in a more rural, laid back area of the city. The first-due area was littered with "supervised care facilities." From what I gathered, this designation really means, "the least crazy resident dispenses the medication."

Just before lunch, we were dispatched to an ill man at an address a couple blocks from the station. We pulled up in front of an old dilapidated house with an overgrown front yard. I pulled the EMS gear off the truck and followed the crew up the porch to the front door. The captain didn't bother knocking; he just walked in with the rest of the gang. The front room looked like nothing I had ever seen. It was a large area, filled dormitory style with beds. Some of the beds sat on frames; others lay right on the floor. A pyramid of four or five television sets sat off in one corner of the room. The highest TV was 6 feet off the ground and couldn't have had a screen bigger than 12 inches. The sets it rested upon had broken picture tubes, missing knobs or some other malady that took their life. A group of residents huddled in front of the altar watching a Flintstones cartoon. I could tell these voyeurs were enchanted by the far-off place and time they could only visit in their dreams.

My captain started calling out for a man named Gilley. One of the residents got up and rushed over. My captain stopped him before he could make contact and said, "Merton, stay the hell away from me and go get Gilley."

Merton was around 5 feet tall and had little doll hands. One of his eyeballs was almost twice the size of the other. He wore nothing but a purple Speedo bathing suit. He had a hair lip, and his nose was crusted with dried snot. He looked like a halibut. I found myself staring so I moved my gaze to the dorm section of the room. None of the beds were made. In fact, most of the beds were filthy mattresses covered with sleeping bags or tattered blankets. Snoring residents occupied several of the beds. A mangy cat, missing an ear and eating a mouse in the corner, didn't seem out of place.

I was snapped back to reality when my captain yelled, "Damn it, Merton, I said stay away from me and go get Gilley!" The rest of the crew laughed uncontrollably as Merton wrapped his body around my captain. My gaze had returned to the cat when the pounding started. It sounded like someone was dribbling a bowling ball in the other room. I went around the corner, toward the noise, in search of the ill man who had summoned our presence to this circus. In the middle of the kitchen, a man sat in a chair. He was bound to a chair by his chest and arms with duct tape. The pounding noise was the product of his bouncing across the wooden floor. He was naked from the waist down. Being brand new at this, I assumed he was the ill man. I went over in an effort to stop the bouncing and take a set of vitals. It quickly became apparent that my new friend was completely unaware of my existence.

I was making a vain attempt to get a blood pressure cuff affixed around the bouncer's arm, but the combination of the duct tape, sweat and perpetual motion made this difficult at best. This was also the first time I had to professionally wrestle with a naked man. A chorus of laughter interrupted my task. I turned to see my new firefighter buddies, Merton and the man called Gilley enjoying the show. Gilley was an older gentleman. He had squinty eyes, large buckteeth and a horse face. He pointed at me and shouted, "Looky thar Jim, that one's playing with Billy. No one ever plays with Billy."

My captain smiled at me and asked, "What in the hell are you doing?"

I replied, "Trying to get a set of vitals on our ill man."

Gilley sucked on his teeth for a second and let loose with, "Son, ain't nothing wrong with Billy, he's just doing his morning hops." I felt like a duck in a room full of peacocks. Merton broke up the scene when he attempted to poodle hump my captain's leg.

Gilley led us to the ill man. He had been comatose under a pile of dirty laundry in the front room. One of the residents had been playing doctor that morning and fed our patient a fruit cocktail of psyche meds. My captain asked Gilley what medicine the ill man had for breakfast.

"Well Jim, I don't know exactly, seems to me there's lots of the big purple ones missing." We performed our standard routine for a drug overdose and transported the patient to the hospital. Gilley and Merton stood on the front porch, waved goodbye and told us they would see us next shift. We drove off, concluding my introduction to the mental-health system.

I was only mildly surprised to see Gilley at our station the next shift. He liked to drop in, have coffee with the "fellars" and keep us updated on his professional life. Gilley mowed lawns for spending money. The fact that Gilley didn't own a lawnmower was putting a major dent in his business prospects. In an act of kindness, the ladder captain gave Gilley an old Sensation lawnmower. Gilley teared up over the gift, even though the old machine required a little work to get it in condition to cut grass. A week or so after Gilley received his gift, I watched him push the mower down the street and into the front of the station.

"Hi Gilley. You get the mower working?"

"Sure did, Pete." (I have no idea why he called me that, and after the third or fourth time I just went with it.) "Stopped by to show Jim."

Gilley was gleaming with pride over the fact that he was able to resurrect the broken grass cutter. He shot me a big smile full of orange teeth.

"Why don't you come back around dinnertime. Captain Jim is studying in his office right now."

It was just after lunch and Captain Jim was taking his afternoon siesta. The captain's office at this particular station had a single entrance/exit, no windows, one desk, six lockers and two beds. Both the engine and ladder captains shared the small room. I watched Gilley disappear around the corner of the station and went back to whatever I was doing. A few minutes later, I was standing in the kitchen when I heard a lawnmower start up somewhere in the station. As I turned around, the door to the captain's office flew open and a conga line of dysfunction shot out into the hall: Gilley and his 5-hp lawnmower running at full throttle, chased by our 58-year-old ladder captain, clad in his white huggy briefs, swearing and vowing painful death to the savant gardener. The carnival headed out the door and down the street into the blinding brightness of the south Phoenix summer afternoon.

You may be asking yourself, what's the point? Why describe a service-delivery encounter that took place over 30 years ago? Most of us began our careers with the romantic notion that the majority of runs would involve saving lives and fighting fire (not necessarily in that order). This is not the case. For every fire we fight and every lifesaving EMS call, we deliver service 100 times to the Gilleys and Mertons of the world. Before I began my career in the fire service, I assumed that some type of institution took care of people not fully (or even partially) equipped to take care of themselves and pilot their way through society. I have since figured out that we are that institution. I don't think anyone in the fire service intended to deliver social services to the disadvantaged; it just kind of worked out that way. Navigating your way through the health-care system when you have health insurance is like falling into an abyss. If you don't have health insurance, you are of no interest to the capitalists who own the system and are left to the mercy of local government. People who lack the resources, will or wherewithal to fight the bureaucracy are left with one option: 911. We have taken this service over by default. We are the only ones who answer their phone calls.

Taking care of these folks is no easy task, but looking back to the crew who introduced me to Gilley and Merton, I think they had it figured out. None of those guys would ever be characterized as social workers or even caring nurturers. They were firefighters who were happiest when they responded to calls where they felt they made a difference. They weren't always overjoyed to respond when the customer's chief complaint was being bombarded by radio transmissions from alien spacecraft, but they never took it out on the client. They patiently listened, calmed the person down and gave them feasible solutions. (Try to wear light-colored cottons and stay out of the middle of the road. Radio waves are always strongest in the middle of the road.)

Delivering a full menu of services to a diverse community will always present challenges. A former president of our local made one of the most poetic organizational statements that I have seen. We were struggling through a period of time, trying to figure out what constituted an emergency worthy of dispatching a fire engine. The city was experiencing a brisk population explosion, and our call volume was rapidly increasing. The number of engines responding to more than 4,000 calls a year was climbing into double digits with a dozen more making more than 3,000 runs a year.

Our fire chief had just written the "Essentials of Fire Department Customer Service." Some members felt we were losing touch with our original mission of protecting the community from the threat of fire and the sudden onset of death. Our department's union president took the time to pen the "Phoenix Fire Department Customer Service Guide." This was a small 25-page guide that basically said our job is to serve the community. If you categorize certain types of calls as disruptions to your day, you're wrong. The city hired us to go on calls. The customer defines the emergency, we don't. This booklet was distributed to every member of the organization at the same exact time. It was stapled to our paychecks.

Chapter 5
A Race Toward Death

I was standing in the checkout line at the QT down the street from Fire Station 30. It was hot out, and I was procuring a bottle of water and a refreshing York Peppermint Patty while my partner, Lynn, waited to pay for his bucket of diet soda. I am convinced this particular gas station/convenience market is the uterus of the universe and the mother of all atomic activity. It is one of the few places on our physical plane where you can watch someone fuel up their half-million dollar Ferrari next to a person putting $3 worth of regular gas into their $200 dollar Chrysler K car.

The QT was doing a brisk business selling gas, nitrate-laden fast food, tobacco products and beer. (Someone once told me the QT franchises in Phoenix sell more beer than all other retail outlets in Phoenix combined.) The clientele had me hypnotized. No fewer than five different languages flew through the air as the three cashiers took cash, ran debit cards and bagged purchases at light speed. A dozen or so emaciated women stood next to the newspaper stand. It looked like the Miss Meth America pageant. The leader of the group was quite tall, and her talent seemed to be scab production. If she had eaten the 40 foot-long hotdogs that spun on the wiener log-ride machine next to the drink fountain, she might have tipped the scales at 90 pounds. She wore soiled blue jeans and no shoes. Her top was a black lace bra. I assumed the hooks on the back that held it in place were broken because the two ends of her lingerie had been tied together in a square knot. The group of them had the vitality of sailors, lost and adrift in a lifeboat on the open ocean, forced to survive on the combination of rainwater and their fellow passengers.

The market looked like the UN. There were a group of lost boys from Sudan, and an older man with heavy whiskers was trying to pay with some type of red paper money from one of the new Eastern European nations. An Indian woman wearing a sari was waiting in line; several people ahead of her stood a slightly intoxicated man wearing overalls and Elvis sunglasses. Among the patchwork of languages in the small market, the predominant dialect was Español. My gaze into the quilt-work of American society was broken by a female voice (speaking English) coming from the radio on Lynn's hip. She summoned our response to a car accident at an intersection half a mile west of where we were currently conducting business. We quickly paid for our provisions and moseyed up the road.

Lynn and I take turns driving our emergency response vehicle. Generally whoever is first to the rig drives. This works out pretty well. During the day Lynn is usually first to the truck, after 10 p.m. I find myself behind the wheel. It's the way our circadian rhythms work. Regardless of which of us drives, we respond Code 2 (no lights or sirens). We don't carry hoses, saws or EMS gear, so unless the sky is red from pyrolysis, there is no need for us to shave 15–30 seconds off our response time. It's kinda like what the old bull told the young bull, "Hell son, let's walk down and screw 'em all."

Even from half a mile away, we ended up the fourth arriving unit. Two vehicles had met in the intersection at an extreme rate of speed. An SUV remained upside down in the middle of the intersection. An older sedan was wrapped around the traffic light 75 feet away. Crews had a total of five patients, two of them triaged as "immediates" requiring extrication. The first in company officer had his hands full with the combination of patient care and acting as the incident commander (IC). The only thing I had in my hands was a bottle of perfectly chilled spring water and a refreshing Peppermint Patty, so I assumed command.

That transfer went like this:

"North Deputy to Command."

"This is Command, go ahead North Deputy."

"I copy you have a total of five patients, the two in the green sedan are immediates and require extrication. I'm going to go ahead and take command and make you treatment sector. Let me know what resources you're going to need."

"Copy, you'll be command, I'm treatment sector. I'll get back to you with resource needs in a few minutes."

This simple command transfer allowed the company officer to focus on the injured customers' treatment and put us in a position where Lynn and I could form an effective strategic level of command, managing resource deployment and the global safety of the patients and incident responders operating in the middle of the busy intersection. The biggest hazard present was the traffic trying to snake by the incident. This group of motorists belonged to the same census tract we just left at the QT. Many of them do not speak English, and I seriously question whether the red, yellow and green traffic signals carry the same meanings in their native countries.

The incident commander is responsible for managing safety for the entire incident scene and in the case of this incident, my biggest safety concern preventing firefighters (and anyone else) from getting hit by a vehicle. My needs were promptly met when the police arrived on the scene. The first officers to get there were the motorcycle cops, followed by patrolmen in police cruisers. This put enough law enforcement on the scene to both investigate the accident and direct traffic. The police quickly began routing traffic away from the firefighters treating the trapped patients. A ladder company arrived and began cutting the green sedan away from its trapped occupants. A minute later a fully turned out firefighter pulled a 50' jump line charged with Class A foam to protect the extricators and patients. I was full of marvel and a mouth of minty freshness as I watched cops direct traffic while firefighters in orange safety vests over blue Phoenix Fire Department T shirts, brush pants and rubber gloves treated the injured peoples. Everything was working like a finely tuned watch; scene safety was at an all time high. Never in a million years did I expect to see what happened next. Remembering it still causes me great agitation.

The sanctity of our scene was broken by a funeral procession. I kid you not. If you or any of your loved ones are associated with the psychotic Barney Fifes who work as funeral-procession motorcycle escorts, please quit reading this and immediately seek the counsel of a mental-health professional. What is it with these guys? Four paramedic engine companies, two truck companies, five ambulances, and about half a dozen cops are working a medical incident across the entirety of the intersection and these misguided lunatics on 1,000cc Kawasaki motorcycles take possession of the scene so a line of vehicles en route to bury a dead person won't be delayed? Am I missing something?

Before I go into the details of what happened next, I would like to take this opportunity to ask you, the reader, a couple of questions. First, what is the hurry to bury a dead person? Most people have been dead for at least several days prior to their graveside service. Why do these fake cops have to race at speeds well above double the legal speed limit to stop traffic for the procession?

Everything about it screams, "We lead the grieving party in a manner so we too can join the dearly departed." After our funeral-convoy-interrupted EMS call was over, Lynn told me about an incident that happened early in his career:

"We were dispatched on a car/motorcycle accident. One of those motorcycle funeral guys was hauling ass, racing to the next intersection to shut down traffic when he hit a truck that was turning. The truck stopped the bike but the rider kept going until he went through the windshield of a car a hundred feet farther up the road. The police estimated the speed when his bike contacted the truck at around 90 mph. He was the first exploded human

I'd seen in my short career. The only way I can describe him was he was whatever the next level is beyond dead. The lady driving the car was shaken up, but other than a few small lacerations from the windshield disintegrating into her, she was OK. I'm sure she twitches every time she sees a long line of vehicles with their headlights on during the day."

I will use my second question to re-ask my first: What is the hurry? Is this ritual an archaic holdover from the days of yore, when we buried the contagious? I could understand if this were an effort to prevent the population from contracting leprosy or the Black Death. But we don't see a whole lot of people dying from diseases best described as the wrath of an angry God, and if that were the case, wouldn't it be prudent to cremate the bodies in some private ceremony, sparing the rest of us the risk of a grizzly ending? Ending these silly antics would certainly increase traffic safety and improve our chances of having an uneventful day. In the end we would also protect these out-of-control motorcycle funeral escorts from themselves. Doesn't the government thrive on protecting people from themselves? I don't know which licensing agency is responsible for allowing this madness. This custom goes beyond ambivalence to insanity. In my book, any law allowing these cop buffs to race around public thoroughfares with reckless abandon is just as wrong and misdirected as trying to legislate relationships between two or more consenting adults. Our government institutions should not sanction anyone (let alone non-emergency responders) to drive motorcycles at near sonic speeds and stop the regular flow of traffic for the dead. I won't pretend to understand what winds up these kamikazes, but when loitering around the mortuary, they look like a group that escaped from a Star Trek convention, got their hands on fake police uniforms and took possession of too much two-wheel horsepower. I urge you to contact your elected officials and demand an end to this silly and hazardous practice.

A few years ago, someone suggested that we put paramedics on old surplus cop motorcycles. In situations where traffic was backed up, a pair of medics on a motorcycle could quickly weave their way to the accident scene. On the surface, this sounds like a completely plausible and sane idea for a duo of gay comic heroes. But it took a full 10 minutes for the assembled group of response chiefs to quit the laughing and disparaging comments before we could answer with a synchronized "no." The final word came from a battalion chief who said, "We ain't turning our medics into a group of those wacko funeral procession guys." This brings us back to our auto accident.

It took me a minute to figure out that the lunatic directing traffic right next to our extrication operation wasn't a member of the Phoenix Police Department. It hit me when I looked over and saw a long line of slowly moving traffic. When his moron partner broke the intersection in excess of 60 mph, I choked on my chocolaty mint. I threw the radio mic to Lynn, left the clarity of the command vehicle and grabbed the first real cop I saw.

"Do you see what those ignorant cocksuckers are doing?" I asked. "Arrest them immediately."

The cop looked over, slowly shook his head, and spoke into radio mic attached to his shoulder. Someone radioed back, and then one of the real motorcycle cops climbed aboard his bike and took off toward the renegade fake cops. In the meantime, the police stopped the remaining quarter mile of funeral goers.

The cop turned back, smiled and asked, "What do you want us to do when we catch the guy?"

"Arrest him."

"For what?"

"Endangering the fire department."

"I don't know if that will stand."

"Then shoot him with your taser until the battery dies."

The cop seriously considered this possibility before opting to write the misguided escort a ticket. It was a solution I would have to live with.

After we finished transporting the last patient (all of them survived the event) and prior to turning the scene over to the tow truck crews, everyone got together to discuss our close call. None of us came up with any reliable way to avoid similar occurrences from happening in the future. It seems as long as people continue to die and their families have funerals for their departed loved ones, we will have to keep a vigilant eye out for the imprudent shepherds of the dead.

Chapter 6
Stick to Fires. Don't Pack Heat.

The fire service has seen enormous change in the last 40 to 50 years—so much change that some have suggested it no longer resembles itself. True, we've gone through our share of metamorphosis, but the big stuff has pretty much stayed the same. You can cover the differences between the last two or three generations with a smooth-bore nozzle (or a fog, take your pick). One constant for the vast majority of firefighters is they are not allowed to carry guns while on duty. This excellent idea has stayed with us ever since Adam of Adam and Eve fame first strapped on a leather helmet. (Everyone incorrectly believes Ben Franklin was man's first fire chief. He was not. Our first five-bugler was none other than earth's first man. An interesting side note to this story is many conservatives, mayors and city managers believe the snake from the Adam and Eve story was the first union president, but that's a different fable.) This excellent rule will constantly apply because firefighters are firefighters regardless of what year it happens to be, and we should not carry guns. The following tome proves my point.

It was an hour before dinner when Captain Jones remembered he had a job to do. Wheeling the green leather office chair back from his steel desk, he got up and headed out of his office in search of Henry, his engineer. Capt. Jones would have hyperventilated and seized had he known that 40 years later, vintage furniture stores would sell the same desk and chair set for a price that amounted to his current annual salary. Capt. Jones found Henry in the kitchen, preparing the evening meal with the two other members of the station. "Henry, traffic is starting to get heavy. It's time." Henry was a tall, strapping man in a John Wayne kind of way. A shit-eating grin broke across Henry's rugged face as he grabbed a large red apple from a bowl of fruit sitting on the pale green Formica countertop. Henry went to his locker to retrieve a blue bandanna and his pack of Lucky Strike cigarettes as Capt. Jones went into his office and pulled a long, soft case from under his bed. He unhooked the leather strap on the end of the satchel, pulled out a lever action rifle and lovingly loaded it with a single round. Both of the firefighters cooking dinner shook their heads as the captain and engineer met up in the day room and walked out the front door.

Forty years ago, Phoenix was a small desert city of around 100,000 people. Station 15 sat just east of the city's busiest intersection. Every weekday at 5 p.m., hundreds of commuters working on the west side of town would pass in front of the red brick, two-bay station on their daily journey home. On this afternoon Henry and Capt. Jones would put on a show for the passing motorists.

Henry went to west end of the front yard, lit a filterless cigarette, blindfolded himself with the bandanna and placed the specially prepared apple atop his head. Capt. Jones took his position on the far end of the 100' front yard across from Henry, then cocked and shouldered his rifle. Westbound motorists, waiting for the light to change, watched in stunned disbelief. When the traffic light west of the station changed to green and the throng of eastbound traffic was in front of the station, Capt. Jones squeezed the trigger, firing off the blank round in his rifle while Henry pulled the string he had attached to the apple, displacing it from his head. These William Tell performances became recurring street theater and caused more than their fair share of vehicular near misses until the day Station 15's battalion chief witnessed the act firsthand and sunset the production.

Firefighters (on-duty firefighters) and guns don't mix. I know for certain if I had a standard-issue sidearm, I would shoot the clocks off the walls of any fire station I may be in. I do not have the discipline or proper psyche to pack heat. Every harebrained scheme that ever attempted to combine firefighters and cops into a single organization has met with failure. I need to look no further than my own fire department for proof. A decade ago we took a group of captains assigned to our investigations division, sent them through cop school and issued them a handgun along with the power to arrest people. In Phoenix, this makes as much sense as training cops to be paramedics. For the most part, our fire investigators are older captains with deeply held pacifist beliefs, some of them borderline Quakers. This silent majority never took their gun out of the box, preferring to leave it in the trunk of their car or on a high shelf in their closet.

This isn't the case for all of the dually sworn fire captain/arson cops. A few years ago I was meeting with the chief that manages our fire investigators. During the course of our conversation, he opened a desk drawer to retrieve some paperwork. I couldn't help but notice several hand guns in the drawer. When he looked back up he noticed me staring at his cabinet full of firepower.

"Most middle managers keep office supplies or candy in their desk. I've even known a few who kept a bottle stashed there, but this is the first time I've seen a chief who can come out of his drawer firing with both hands," I said.

"Anytime you want to switch jobs, feel free," he replied. "I'll take 400 unarmed B-shifters over a couple of arson investigators who sleep with a picture of Dirty Harry on their nightstand. This city is a much safer place with their guns in my desk drawer. As far as how they ended up in here, I'm not at liberty to say."

A decade ago our city and fire department enacted a rule prohibiting employees (with the exception of uniformed crime fighters) to possess a gun on city property. With the exception of a few 2nd amendment zealots, the policy was met with a yawn. During the ensuing 10 years, this policy has had no adverse effect on our mission as a fire department. On the other hand, it has probably prevented stupid, and possibly lethal, accidents from taking place. Like most well-thought-out policies, this one had been informally enforced for quite a while in many fire stations prior to the formal adoption of the "no guns at work" rule. The following story illustrates the infinite wisdom behind this rule and its liberal application.

It was the mid-1980s, and I had been a captain for less than a month when I roved into an engine company where a guy nicknamed Copperhead was the senior firefighter. At the start of our shift, we exchanged the standard morning pleasantries. As Copperhead walked past me, I noticed he had a gun stuffed in the waist of his uniform pants. The ink may not have been dry on my captain's shirts, but I knew that on-duty firefighters (Copperhead in particular) shouldn't carry guns.

"I predict a pretty low-key shift. You won't be needing the gun, Copperhead," I said.

"I'm carrying it for protection from my enemies," he replied. "I feel safer with it."

The rest of the conversation didn't last 2 minutes, and I safely stowed the gun for the remainder of the shift. Copperhead was a very distinct individual, and I really wasn't interested in a long soliloquy about his enemies, real or imagined. All I cared about was protecting myself, the rest of the crew and the general public from a firefighter whom I barely trusted to run a lawnmower. At the time, I felt Copperhead's overactive imagination verged on neurotic. Eighteen years later, a group of us were reminiscing after lunch when I learned the true source of Copperhead's cocked-and-locked approach. It is one of the greatest lunchtime stories ever told.

Copperhead's family had owned a mining claim in the mountains north of Phoenix. I don't know if it was the allure of untold buried riches or another petty issue that can cause a family to choose up sides and go to war, but Copperhead's clan is the last one in our state's recorded history to have a shoot out over a worthless piece of high desert with a hole in it. The entire family unit was rugged outdoor types. The feud was officially underway when the half of the family ending up with the spread (Clan A) had the other half (Clan B) evicted by local law enforcement. Clan A felt they could double the profits on all of the precious ore under their feet; all they had to do was get it out of the ground. Clan B was bitter and determined their no-count relations were not going to keep them from their share of the underground wealth. The dysfunction went public after Clan B procured an airplane.

Clan B loaded up their small plane with all variety of rifles and handguns, removed its doors to provide a clear shooting alley, and circled the property, firing at anything and everything that moved. Sometimes they would do this late at night to disrupt the sleep of their mining kin. Clan A's mining operation was severely interrupted by the ousted clan's "death from above" antics. This circus occurred frequently enough that it drew the attention of the sheriff and the media.

Living in Arizona, I paid little attention to the televised media accounts (this type of behavior is a normal occurrence here, generally reported on the local news between the weather and the sports). I never made the association that these were Copperhead's people. Had I known, it wouldn't have mattered; he wasn't going to be armed in a station with me.

According to lore, the gun battles (including the aerial ones) stretched from months into years. This took place in a very isolated area where the few residents pretty much keep to themselves. The sparse law enforcement presence had thousands of miles of county to police and didn't have time to baby-sit the family squabble. Unless Copperhead's dueling clans were caught red handed, shot an innocent bystander or a spotted owl, there wasn't much to do. Mining continued and each day brought the promise of striking the mother lode. Not wanting to leave the property for fear of outsiders or shunned relatives showing up to steal the unearthed riches, Clan A made the decision to become self-sustaining. After digging a well, they went out and purchased an adolescent bull buffalo to meet their long-range protein requirements. The A Clan mined, and the B Clan continued to harass them while the buffalo grew to enormous proportions. Once the animal tipped the scales at 2,000 lbs., Clan A decided to turn the beast into hamburger and T-bones. A doublewide horse trailer was selected as the slaughterhouse. The center divider was removed and they spent the next few hours coaxing the brute into the steel trailer.

Grazing animals have always lured us into a false sense of physical superiority. We eat meat therefore we are to be feared. We are smarter and have thumbs. Buffaloes are related to cows; teenage boys in the Midwest tip cows over at midnight. Seems reasonable that the fearsome, intelligent, thumbed human should be able to best a stupid buffalo. I am told by associates who hunt that adult buffaloes have no natural predators. Buffaloes are not friendly animals. They are mean-spirited, powerful and fast as lightening—nothing any other animal wants to mess with. What happened next really shouldn't come as a shock. Once the buffalo made its way to the yummy hay at the front of the horse trailer, the ramp was lifted and locked in place, then Clan A's senior male shot the animal with a 45-caliber handgun. This only served to infuriate the one-ton beast.

The clan stood in slack-jawed amazement as the wooly one-tonner bucked, kicked, and used its large head and neck like a battering ram, completely dismantling the steel horse trailer like it was made out of cardboard. It happened so fast that the tribe didn't have time to react. Before they knew it, Mr. Buffalo was free and started chasing the family. It took a while but the clan all made it back to the relative safety of the main house. Everyone's anxiety levels went through the roof when the buffalo started ramming its head into the house. Windows shattered, doors were busted and the posts holding up overhangs and patios were splintered as the clan huddled in fear. The only things that kept the charging animal from breaching through the front of the house were the large sheets of collapsed roof covering the window and door openings.

Every time one of the clan attempted to escape from the house, Cujo on steroids would charge. The wounded animal besieged the Copperhead clan for six full days and nights. The buffalo held the family hostage and wanted payback. Whenever the crazed animal saw signs of life from the plagued dwelling, he would run full speed (which is in the neighborhood of 40 m.p.h.) and launch his wheelbarrow-size head directly into the building. On day two, the situation became desperate when the aviator clan circled overhead and pumped hot lead through the roof. The rain of gunfire renewed the buffalo's frenzy. He repeatedly rammed the family car, pushing it 50 feet and over the ledge of a steep ravine. The next few days were a waiting game. By the fourth day, Clan A had all but given up hope of leaving their precious claim. They were given back their freedom on day six, when the buffalo finally succumbed to its gunshot wounds and died. The death of the large beast was both a gift and an omen. The weeklong barrage, compliments of a one-ton, rabid freight train, had soured life on the quarry. Within weeks the mining operation was abandoned, and the clans dispersed on the dry, hot desert winds.

I believe there is a lesson in this for all of us. Most families do not take their dysfunction to the same level as the Copperhead clan. This is because most families don't shoot at one another (or a one-ton devil beast). The band Lynyrd Skynyrd said it best: "Handguns were made for killing; they ain't good for nothing else."

Chapter 7
False Imprisonment

I would like to preface this essay by stating I really like the police. I work at a firehouse where it is common to have as many cops in residence as there are firefighters (there are 12 of us). It is odd to wake up at 3 a.m. and not run into at least half a dozen Phoenix police officers sitting in the dayroom, eating fast food and telling stories with the TV blaring. They are nice people and do a job that I don't think I could make a career of.

As a general rule, firefighters don't enforce anything. Our customers call us when the genie of mayhem and destruction comes to visit. We arrive and beat the sprite back into the bottle. We are the good guys. On the other hand, the police exist to enforce the rules we call laws. This responsibility comes with the authority to take people to jail and—in certain situations—to shoot them to death. This is the reason they will never be as popular as we are. The following incident beautifully illustrates my point.

The four of us were taking our time as we drove to the call dispatched as an unknown medical. It was a few minutes after 1 a.m. when we pulled up in front of a small house that sat in the center of a dozen square miles of urban blight. The front door of the house stood wide open and threw light into the warm night. It had been a very busy day; this was our 11th or 12th or 15th call. We didn't know or care; we quit counting hours earlier. Mama and papa bear greeted the four of us with nods and dull stares. On the sofa, in their jammies, sat two little baby bears. We greeted them: "We're with the fire department. How are you folks doing this evening?"

It quickly became apparent that neither of the parental units spoke Ladder 11's common, singular language. I turned to the kids and asked if they could verbally communicate with us. The older boy child replied, "I suppose so." Super. "What can we do for you?"

The family began to communicate in a tongue Ladder 11 could identify but not comprehend.

This went on for what seemed like a week then the boy turned to us, came up for air and said, "My older brother came home a couple hours ago and went into the bathroom. A few minutes ago my mom went in to see what he was doing. He is in the bathtub. My brother told my mother to get out of the bathroom or he would hit her. My mother said that my brother shouldn't be doing that to himself in our bathroom. My brother drinks and uses drugs."

Every single call had gone exactly like this all day long. One of our engineers made balloon animals out of rubber gloves for baby sister, I continued to interview little brother, and a firefighter and engineer No. 2 headed down the hallway toward the bathroom to discover what unknown medical lie on the other side of the door. I couldn't help but peek behind the curtain when I heard Rodney the engineer bark through the open door saying, "If you keep that up you're going to need glasses."

One did not need to speak two languages to figure out what big brother was doing. He was spread out in the tub wearing nothing but a look of crazed intensity. It was obvious that he had very good circulation and profusion to all of his body parts. He also had good muscle control. Rodney the engineer was right; if he kept it up he was going to need glasses. It took a few minutes for big brother to notice he had an audience of highly trained medical professionals. This seemed to upset him, but he didn't stop his fast-twitch activity as he began the long climb out of the tub. As we stepped out of the bathroom, big brother settled back into the tub and his furious rhythm. Once you add everything up, deciding what to do next was a no-brainer. Big brother was most likely under the spell of some type of intoxicating substance. He had been at it for two hours. His airway, breathing and circulation (the classic ABCs of the EMS world) were top notch. He wasn't bleeding, and all of his bones and soft tissue seemed to be intact and functioning properly (albeit inappropriately). None of us felt that big brother needed medical intervention.

The group of us had no intention of wrestling with a high, fully aroused, soaking-wet adult male. We told the family (via little brother) to keep an eye on big brother and call us back if they needed us. Everyone smiled at one another and we headed out the door. As we got to our rig several police cars pulled up. I was briefing one of the officers when he interrupted, "Doesn't the fire department have hair regulations? You and your engineer have hair like a girl." It was true; we both had long, luxurious hair. Mine had a loose, natural curl while the engineer's was the product of a chemically induced

53

perm. Both of our hairdos fell well below our shoulders, covering up the center of "Phoenix" on the back of our uniform shirts. I replied to the snitty officer, "No, in fact we don't have any regulations at all on the Phoenix Fire Department. If you want to see a shorthaired manly-man go in that house and check out the guy in the bathroom. He's waiting for you." With that, Ladder 11 drove off into the night.

We hadn't been back to the station for 10 minutes when the lights came back on, accompanied by the annoying tone. The voice dispatched us back to the house we had just left to "assist PD." We were greeted by complete pandemonium when we pulled in front of the house the second time. Papa, mama and the baby bears were trying to shoo the police away. The center of this extravaganza featured the cop who had lectured me about my hair. He was attempting to perform police procedures on big brother. Big brother and the cop were rolling around on the ground, both covered with water, dust and mud. From the front seat of the truck I could see that big brother was still nicely profused and heading to jail for the evening.

Two different public safety agencies—same call—drastically different outcomes. The fire department takes the approach that if our intervention can't make the situation better we will not make it worse. In many cases all this requires of us is we show up, observe, decide, inform and leave, which we did that night. Our department lives by the credo of Winston Churchill, who said, "If you are going to shoot a man, it doesn't cost anything extra to be nice about it." Although being nice to your customer doesn't cost a dime or take any extra time, it pays gigantic dividends. People will generally like you more when you are nice to them.

After they hauled big brother off I asked one of my cop buddies at the scene what had happened after we left. He told me Officer Hair Nazi went into the bathroom and ordered big brother to knock it off and answer his questions. Big brother came out of the tub with a handful of himself and bad intent. This earned big brother a face full of pepper spray and a ride to the lock up. As my cop buddy finished his story, he rolled his eyes and shrugged his shoulders. We both knew the whole episode didn't need to happen. It isn't fair that one misdirected individual can discolor an entire profession of decent, hard-working people. It bears repeating: Most of the cops I come in contact with are professional, helpful and kind. I have worked with firefighters that made Hair Nazi Cop look like Captain Stubing from the popular TV show, "Love Boat." I believe these individuals are the exception (in both of our organizations). The problem is they can do tremendous damage. It only takes a single out-of-control professional basketball player to go after a fan and wreck an entire season.

I had the misfortune of being on duty Nov. 29, 2004, and seeing this happen first hand between the Phoenix Fire Department and the Arizona Department of Public Safety (our state's highway patrol). I have read accounts of law enforcement officers arresting firefighters because they didn't like where they parked, but it had never happened in my neck of the woods.

I always get a sense of deep, personal dread when I get urgent messages to call someone immediately. My first thought is that someone I love has been maimed or killed. This message began as a page from our Alarm Room that read, "Call alarm ASAP." My paranoia concerning urgent "Call ASAP" messages has never been validated. Every one of these imperative "Call Right Now!" requests has proven to be notification for some impossibly idiotic, not life-threatening incident. As is always the case with these types of situations, my partner Lynn and I were sitting down to eat lunch and our food had just arrived. Before I could get the cell phone out of my pocket and dial the seven numbers connecting me to Alarm, the hostess came to our table and asked, "Are either of you Nick Brunacini?"

With a deepening sense of dread I nodded and told her, "Yes."

"You have an urgent phone call."

I got up and marched toward the phone, "Hello, this is Nick."

"Hi Nick, this is Shirley in Alarm. We have a situation on the freeway. A few minutes ago one of the crew of Engine 41 cleared us over their tactical channel and requested a BC because DPS (the Arizona Department of Public Safety) had just arrested their captain."

"They were on a call?"

"Yeah, evidently the officer wanted 41 to move their rig and arrested the captain for not complying. Battalion

4 is on the scene now and wants a shift commander to respond."

As I hung up the phone I felt relieved that no family members or on-duty firefighters had been blown up in a freak accident at a high-school chemistry lab. Lynn and I dropped $25 on the table for a pair of lunches that would go uneaten and headed up the road. On our way to the unlawful arrest of our company officer we called BC4 on the cell phone and learned that the DPS had the sense to move the operation off the freeway and behind a convenience market a couple of blocks east of the interstate.

After 10 minutes of driving and quiet reflection, we made it to the scene of the incarceration. Three DPS cruisers were parked in the corner of the lot, behind the store. Battalion 4's rig sat behind a new 1,500-gpm pumper on the street bordering the store. We pulled into the parking lot, and I as I got out of our response vehicle I could see our captain sitting in the back of one of the cop cars. I looked around in search of hidden cameras, still in disbelief that this was happening, feeling that it was possible this was a hoax, maybe a set up for one of those insipid reality TV shows. As with most of life's mysteries, fact is stranger than fiction.

A very tense silence embraced the scene. The three DPS officers had the simmering posture of parents who had just dished out stern discipline to their misbehaving children. The crew of Engine 41 sat on the curb, spitting sunflower seed shells into the street. They all had the look of resigned disgust on their faces. The BC and his driver just stood there shaking their heads. The only participant who appeared "normal" was the captain who was sitting handcuffed in the back of one of the DPS cruisers. He smiled and nodded as I walked up.

Not knowing what else to say, I blurted, "Why is our captain handcuffed in the back of one of your cars?" This was answered (by both a sergeant and lieutenant) with something to the effect of, "We will explain." But, "We will explain" quickly turned into a sermon: "We are sick and tired of how you guys operate on our freeways."

I really wasn't looking for a lecture about our operational inadequacies; I still couldn't believe that the Arizona Department of Public Safety had arrested one of our on-duty members for simply doing his job. In the interest of economy and finding out the uneditorialized version of events, I walked over and asked 41's crew what happened. The engineer quickly became the spokesman for the crew.

"We had just gotten back in quarters from a traffic accident on the freeway when we got dispatched to another accident in the same location. We arrived to the scene and found a couple of cars parked in the emergency lane on the right side of the freeway with people in each car. The cars were 3 or 4 feet away from high-speed traffic in the right lane so we parked just behind the first car and took the right lane (effectively turning a three-lane freeway into a two-lane freeway).

The captain and firefighters were evaluating the people in both vehicles when a highway patrol officer approached me. The DPS officer came over and told me to move my rig up ahead of the accident into the emergency lane. I told him our department's policy is to park our rigs in a manner that protects us and the patients from traffic. We couldn't open the doors of the accident vehicles and get into them without stepping into oncoming traffic, so I told the DPS officer if he wanted my truck moved he would have to talk to my captain. You can see how well that worked out. After he arrested our captain he came back over and told me that he was now in charge of the scene and to move the truck. I told him if he wanted the truck moved he would have to move it himself. He climbed in and moved it."

I stood there in awe that someone who had done something so bone-headed and stupid had the intelligence and technical ability to drive a fire truck. The world is a strange and far-out place.

I went back over to the assembled group surrounding our detained captain. My partner Lynn was talking to the DPS officer who made the arrest while sarge and lieutenant looked on.

Lynn asked officer Barney Fife, "Was the captain rude to you?"

"No. Your captain was very polite, he was a gentleman but he disobeyed a direct order from me. I gave him another opportunity to comply. I arrested him after he disobeyed the second time."

The arresting DPS officer looked like he was 16 years old. I couldn't take anymore and blurted out, "OK that's enough. This is complete bullshit. There are hundreds of people selling methamphetamine to children and you assholes take time out of your busy day to arrest one of our captains for operating according to our department's safety procedures. If he were a motorist you pulled over for going 90 miles an hour you would have written him a ticket and let him go. Let him out of the back of your stupid car and take the handcuffs off of him right now."

This outburst brought on the beginnings of another lecture about us screwing up traffic on their freeways. I couldn't help myself. "I refuse to stand here and listen to a safety lecture from you. The last half-dozen of your guys that died in the line of duty were killed by traffic. This thing has evolved way beyond you being pissed off about us stopping traffic. Let our guy out of your god damn car, now!"

One should not curse and yell at people who carry guns. The DPS officers made the standard gesture one would expect when challenging sworn officers of the law. They took a step back, extended their arm straight out with their hand up in a stop signal while their other hand naturally dropped to the gun on their hips. Walking away was the smartest thing I did all month. I went around to the front of the store to call my handlers and let them know what was going on.

While I was on the phone one of my counterparts showed up and took over the impossible task of communicating with the Arizona Department of Public Safety representatives. He was more calm and lucid; when I returned to the group our captain had been released from the back of the car and decuffed. The arresting

DPS officer had left the scene and we concluded our conversation with his sergeant and lieutenant by thanking them for creating hundreds of future meetings between our organizations.

The actions of the Arizona Department of Public Safety were stupid, negligent and mean. I have looked at the issue (as best I can) from their perspective. I understand that the main part of their job is keeping the freeways flowing and clear but doesn't the title "Department of Public Safety" imply a little more than "let's keep this traffic moving"? It is true that a 16-ton fire truck, purposely parked in a way that blocks a lane, will have an adverse effect on the smooth and orderly flow of traffic. The same thing can be said about an accident that occurs on the roadway. The difference between the two is traffic will continue to whiz by the accident, it won't whiz by a fire truck that has been strategically positioned to protect both the rescuers (including the cops) and the accident victims.

Our safe parking practices have saved us from countless secondary collisions. In many instances had our apparatus not been parked in a way that protected the scene the later arriving crashers would have driven their vehicles right over the top of us. Vehicle accidents continue to be a leading cause of death for both police and firefighters.

We ain't changing our parking policy. If we have to work in the street (a high-hazard area) we will use whatever means necessary to protect ourselves. In many cases this is accomplished by using our apparatus as a shield from oncoming traffic.

In the end, our little fiesta with the Arizona Department of Public Safety wasn't about how we park, the orderly flow of traffic, or miscommunications between an engine crew and a highway patrol officer. I still have trouble believing that they arrested one of our members and kept him handcuffed in the back of his car for almost an hour. The only plausible explanation I have been able to derive from this event is it was about an immature cop who wanted to show the world he was in charge.

In the hours following this event we added the response of a command officer to any incident that occurs on a freeway. During the course of the ensuing week, our captains and battalion chiefs have been overwhelmed by responses from members of the highway patrol. This is because the vast majority of the Arizona Department of Public Safety officers are decent, competent and nice. Most of the DPS officers we've interacted with in the subsequent days were more upset and embarrassed that the incident happened than we were.

Life goes on and firefighters and cops will continue to respond together to accidents that occur on the freeways. The outcomes of these incidents will be affected, to a large degree, by how well we get along with one another.

Chapter 8
Young & Old

I am awash in deep reflection. I just read a highly classified internal memo describing how the Phoenix Fire Department is going to hire 250–300 new firefighters in the next several years. This number is more than triple our typical rate of new hires. Large numbers of retiring members coupled with wild growth in the Valley of the Sun have created a hiring boom. In the same load of interdepartmental mail, I also received my 25-year service pin. It's a dazzling fake gold Phoenix bird with cubic zirconium eyes. In some twisted Dilbertian way the yellow bird pin completes me. The 25 years puts my career more than two-thirds behind me, causing a certain sense of serenity, much like a 12-year-old boy in an idyllic alpine world, sledding with his wet nurse and pet wolf dog. This thought and these feelings compete with my alter ego: a deranged, one-eyed bag lady who collects headless dolls. This is what a career in the fire service does to a person.

During the late 1970s and early '80s the Phoenix Fire Department hired a large number of firefighters. Our training academy was designed to accommodate 25 recruits. Hiring was so out of hand that we had to cap academy-class sizes at 40 recruits. This trend continued into the mid-'80s when the fire department took over emergency ambulance service. After the hiring bang of the mid-'80s, hiring dropped way off. I estimate we've hired an average of 30 or 40 new firefighters a year during the past 20 years. This has caused us to age as an organization. In 1985 our department's average age hovered in the mid-20s. We were young, full of piss and vinegar, and didn't have a care in the world. Today, almost half of the Phoenix Fire Department requires the aide of reading glasses and is eligible for retirement.

I would like to take this opportunity to send a shout out to all of the new firefighters just beginning their careers. The majority of your first year will be devoted to learning the basics: hose lays; the proper techniques for throwing ground ladders; the fact that emergency responders should not shave expectant mothers "down there" prior to delivering babies; and all of the other essential, front-end training that goes into learning how to be a firefighter. I don't want to talk about any of that stuff. Instead, I would like to discuss fire-service culture and how you fit in, along with some explanations that may help you better understand our closed society and freemason-like ways.

The ancient Greeks established what has become the modern fire service. This same group also invented the multilevel marketing system (i.e. pyramid or Ponzi scheme), so everything was designed around seniority. The oldest and wisest Greek fire chiefs would occupy the highest chair in the room. They proudly wore togas of spun gold while the more junior members of the fire company mowed the grass, cleaned the bathrooms and did busy work. Prior to any type of career-development activity, the new guys were required to rub soothing emollients (made by Amway) into the skin of the older men. As newer members were added to the company, the junior members would move up in the pecking order. This level of seniority allowed the sophomore firefighters to wrestle and play ring toss with the older firefighters. Another added bonus for the older members were the halftime festivities, where the probationary firefighters gave sponge baths and fed grapes to the senior members of the fire company. Over the eons, our physical fitness programs, apparatus and uniforms have changed but we still cling to the organizational and cultural systems developed by our Greek ancestors.

Our most significant and profound cultural changes have occurred within the last 50 or 60 years. Nowadays, we select firefighters based on their qualifications, not the color of their skin, ethnicity, gender or sexual preference (a very big hitter in the ancient Greek fire service). Another major cultural advancement took place in the late 1970s when the forward thinkers of our service added emergency medical response to our service delivery menu. Taking on EMS blew a big hole in the "We will Never Change" bunker. This simple and profound shift slowly brought about the sunset of the good old days when we admired the size of our gonads by the firelight (a carry over from our Greek fathers). Prior to the dreaded EMS, the barometer of being a good firefighter was your ability to press the attack and take the pain.

Today life moves a lot faster and some days it seems like our service no longer resembles itself but at our core (and culture) we are still who we are. One of the constants within our organization is an obsession with trying to figure out the most recent generation of new hires. I just sat through a two-day seminar where one of the classes was titled, "Understanding the New Generation of Firefighters." I am sure it will amaze the new generation to find out the biggest difference between you and anyone born before 1980 is your ability to use a computer and the amount of talk time you spend on the cell phone. You also value your time off more than working overtime (this will change when you obtain a mortgage).

Another flash to the blindingly obvious is that you young whippersnappers like your information MTV style—in short, intense bursts. You mean you don't like long boring lectures that seem to never end? The nerve. The problem with accommodating you on this issue is it forces us to redesign training programs, some of which have worked just fine for the last 50 years.

Some of the seasoned veterans will tell you this organizational need to understand you is based on the need to assimilate you more completely. Don't fall for it. Sometimes the ulterior motive behind the dissection of the latest generation is to determine their shortcomings and weaknesses so they can be used to beat you over the head. This accomplishes two things: 1) It tends to make you question your self worth; and 2) It makes the beaters feel better about themselves. This behavior is often justified as "testing the mettle" of the young employee. This boorish conduct also gets packaged as your "right of passage." When this cycle repeats itself, it becomes the gift each previous generation of miscreants gives the next. If you follow ugly children home you are most likely to find ugly parents.

Many people who use the term "right of passage" use it incorrectly and are generally dildos. Right of passage means that you have worked to achieve a goal and have now earned the rights that go along with that accomplishment. Right of passage is sometimes bastardized as an excuse to lord over and treat new members less than nice. It is disturbing to watch a three-year "veteran" act with an arrogance toward junior members that one would normally expect from a table full of super models being rude to the restaurant staff. It is the responsibility of the senior members of the group to bitch slap these prima donnas down and restore organizational harmony. Nothing will wreck a fire department any more thoroughly or quickly than internal fighting and nastiness.

How an organization treats its young says more about it than any other single thing. You hired on as a firefighter and should expect to work hard and get dirty. You didn't sign up to be an emotional piñata for broken and angry employees to take their frustration out on.

As with most things, the truth lies somewhere in the middle. New people shouldn't be treated like houseboys or chambermaids. On the other hand don't expect the group to throw a party every time you enter the room. We expect things of the new folks. Keep in mind that most of your workmates (particularly old crews) have been together for a long time, developed deep meaningful relationships and are loyal and protective of one another. You are new and everyone is forming an opinion of you.

A good rule of thumb for new firefighters is whenever anyone is working, you should be working. You shouldn't have to be told to do basic things like throwing out the trash and cleaning the bathroom. Expecting a probationary firefighter to make coffee is not abuse. It is bad posture for the new guy or gal to fall asleep in a recliner after lunch while the rest of the crew does the dishes and cleans the kitchen. You should do a little more than the older members.

It is important to keep in mind that the world turns in circles. You are just beginning your career. Before you know it you will be taking promotional exams and moving up the ranks. A few weeks ago one of my workmates told me about a conversation he had with one of the guys that worked in his battalion. This member came to him because his son was testing for the fire department, and he was concerned about one of the captains who sat on the final selection board. Here it is in a nutshell:

Member: "I'm worried that my kid won't get a fair shake because Sam is on the final selection committee."

BC: "Sam's a good guy, why would you be worried about him?

Member: "Because he was a booter (probationary firefighter) at our station and I wasn't impressed with him."

BC: "And?"

Member: "Well, I wasn't always as nice as I should have been."

This is a classic example of the chickens coming home to roost. Someone decided they could be mean to someone else. At the time the older member had the power and the new guy just had to take it. A decade or so later the member is highly stressed that there might be major repercussions for his nastiness. As luck would have it, the member's offspring dropped out of the testing process and moved to another state to get away from his father. A few months later Sam took the vacant captain's spot on the member's engine company and soon thereafter (actually less than 24 hours later) the member left the station and is now roving on another shift. Sometimes life just works out like it's supposed to, which leads us to the major difference between the fire service and the rest of today's employment world.

The single biggest cultural difference between the fire service and every other profession is lifetime employment. When I began my career in the fire service it seemed that many of the other large companies filled their work rosters with career employees. Companies like IBM, the big automakers and General Electric—along with the rest of corporate America—had fulltime, lifetime workforces. Lifelong employment meant a person could start with an organization and work their way up the ladder. They had a real pension, insurance, earned leave and could raise and support a family.

This is not the case in today's employment world. Most workers will work for seven or more different companies during their working life. Defined benefit pensions are rapidly being replaced by 401Ks, and employers and their workforce are struggling with ways to keep up with the skyrocketing cost of health insurance. The culture of the American workforce has undergone massive changes over the past 20 years. The vast majority of new firefighters go to work in a place where they will spend their entire professional career. Job security is a good thing and has had made our entrance process much more competitive. You know the world has turned upside down when lawyers are signing up to take a fire department's entrance exam.

The fire service's cradle-to-grave employment practices can create what some city administrators have termed secret societies. What this means for you, Mr. and Ms. Young Firefighter, is you have joined a place where everyone knows everyone else and is suspect of anyone from outside the group. Another cultural reality is everyone on the fire department is somehow connected to one another. I don't mean this in the "all for one and one for all" sense; I mean it literally. Everyone within a fire department is related through blood, marriage, divorce, fishing or any of the other countless human activities. It is not smart to speak ill of a workmate (whether or not they deserve it) in the company of other firefighters.

I remember a story an old engineer told me when I was just beginning my career:

"We had just finished lunch when a woman was walking up to the station. She was an older gal and a little on the heavy side. One of the regular crewmembers was on vacation that day and had been replaced by a young roving firefighter. As the kid watched the lady make her way toward our front door, he said, 'I hope she doesn't want a blood pressure, because we don't have a BP cuff that big.' Before he could finish giggling over his witty remark the captain replied, 'I don't think the lady wants her blood pressure taken; she's here to see her husband.' It took that kid 10 years to recover from that comment. Son, you can be here your whole life and still not know who's hooked up to who."

Another significant difference between the firefighter's job and most other occupations is the fact that it is pretty easy to get killed when you operate inside buildings that are on fire. As if this isn't bad enough, we routinely operate at non-fire emergency scenes where you can die. Because the older members of the organization have been doing this kind of work a lot longer than you, it would be wise to learn the behaviors that have kept them alive and allowed them to grow older. It is impossible to sit through a modern-day tactics class without hearing the instructor refer to the "set of slides" contained in the heads of the seasoned veterans.

It is the incumbents' job to pass what they know on to the next generation. This is the easy end of the deal. The young need to take all of that training, skill and knowledge, make sense of it and apply it in the worst possible work place one could imagine (a building on fire). As stupid as this may sound, this is the easy part. The bigger challenge happens after you have been around a while. It is your responsibility to take what you have learned, mix it with your experience and figure out how to elevate your departments operations and service delivery to the next level. This is the challenge. The only way things get better is when someone figures out a better way of doing business.

Today's firefighter faces a much different workscape than any that came before them. The world is a much more hectic, fast-paced and hazardous place than it was just a few years ago. The incumbents worry that the new generation isn't ready. I say phooey. My youngest daughter is a 10-year-old fifth grader, and she generally misses the homework math problems I help her with. The newest generation of firefighter is smarter, stronger, prettier and better smelling than their older counterparts. Give them 10 years, and they will be every bit as grizzled. Let me be the first to give them a big Dating Game kiss.

Chapter 9
Family Ties

All of our department's apparatus bear the verbiage, "Our family helping your family" in faux gold leaf. This statement is true because the members of the Phoenix Fire Department operate more like a giant family than your typical run-of-the-mill large corporation. I've heard it suggested—on more than one occasion—that we change our motto to read, "Our dysfunctional family helping your dysfunctional family." This wasn't meant in a bad way, but some days it seems as if we're part of an extended episode of the "Jerry Springer Show." It's my observation that this family-style atmosphere exists in most fire departments.

Like everything else in life, family requires balance; this is especially true when you belong to multiple families. Balance becomes even more critical when you factor in all of the families we help (our customers). The members of most professions don't make heavy emotional investments in their customers. This isn't a judgment, merely an observation. When was the last time you had a significant emotional experience with your banker, mechanic, appliance repairperson or dentist? Most service-delivery encounters shouldn't come with a whole lot of emotional baggage because most service delivery tends to be technical and clinical. Can you imagine if the simple act of having the oil in your car changed moved you to tears? Living in a way where you go from one profound, life-changing event to the next leads to lots of really long days, eventually requiring a prolonged stay at the asylum.

High emotion is one characteristic of our work. In fact, we have one of the most emotionally charged careers in the world. Anytime serious emotion, death, or mayhem happens in the community someone calls 911, and we show up. Oprah's got nothing on us. If there were more of a who-done-it aspect to our profession, we would end up with more stupid firefighter TV shows than there are stupid cop shows. Thankfully this isn't the case. A lot has been written and said about how our work has changed over the last several decades. Although the details of our work have changed, the types of calls we respond to and the service we deliver have changed very little. The fire department remains a very close-knit group that delivers an entire spectrum of services to customers who fall somewhere between master of the universe and human fodder.

It's common to sit in fire-department meetings and listen to workmates theorize about the feasibility of creating a "third" service that responds to and handles all the calls that feature the blind, crippled and crazy. These long (and usually emotional) speeches are based on the assumption that these "social" calls have skyrocketed in the last few years. The last time I sat in a meeting dominated by this dialogue, I daydreamed myself back to a mythical time 15 years ago when all we did was slay dragons and rescue seriously distressed maidens with ample bosoms. I call this a mythical era because it is every bit the myth today that it was back then.

Back in the days of old, I served as the captain on a

BLS truck housed on the east side of downtown in Fire Station 11. This is where my third family cohabitated and was where I perfected the fine art of staring directly into the sun. (I call this my third family because my first is the one I was born into, and my second is the one I created through a marriage. The third is the one where I work and have spent 33.3 percent of my life. The third family continues to provide pay and benefits to support the second family.) Station 11 houses an ALS engine, a truck company and an ambulance. During my 10 years there, the station averaged more than 10,000 runs a year. Most of these runs had nothing to do with dragon slaying and rescuing amply bosomed damsels. All of these runs were visits to my fourth family—the one we advertise helping on the sides of all of our rigs. This big family loop is becoming rather incestuous (it's starting to sound like the linage of European Royals), so let's move on.

Some members of our fourth family included regulars who called us a couple times a month. They lived in an old hotel that had been converted into a very basic assisted-living center. Tom and Mary were husband and wife, and they were both legally blind. Tom worked outside the facility while Mary had a part-time gig she did at home. The emergency that always summoned us to their apartment was Mary's premonition that she was about to have a seizure. I saw Mary a couple times a month for five years, and she never had a seizure or any other medical malady that we could diagnose. After the third or fourth encounter, we began to put the pieces of the puzzle together.

Mary's call for help always brought her downstairs neighbor man to her door prior to our arrival. Initially we assumed this was Mary's husband. Several incidents into our relationship, a gentleman with a white-tipped cane showed up and announced himself as Mary's significant other. This left us wondering how Mary's downstairs neighbor knew about Mary's inaccurate seizure aura before we made the 3-minute trip from our station to her domicile. Mary's apartment always had the bouquet of consummated genital congress; it smelled like a house full of teenagers whose parents are on an extended vacation. It seemed that Mary and her downstairs neighbor were engaging in the game of touchy-feely while her husband was away. Mary's crooked or missing clothes never resulted from epilepsy-related seizures, and her concerned neighbor's fine sheen of sweat was never the product of physically restraining Mary from swallowing her tongue. We had become their safety valve whenever the illicit couple had the hunch that hubby was making an early trip up the stairs. We are not private investigators, and none of us ever felt it was our duty to host an intervention for the love triangle.

Over the ensuing five years, we continued to pay our bimonthly visits to Tom, Mary and the concerned downstairs neighbor. I can't recall ever using a piece of medical equipment or transporting anyone to the hospital, but our little Ladder 11 family learned all about Tom and Mary's clan. Tom's mother-in-law never cared for Mary (talk about a premonition) and felt that two blind people shouldn't wed. Mary kept a large organ in her small place and our bond grew to the point where one of our engineers would play old standards while Mary sang. She had a beautiful voice.

Early in our relationship, we went on a lunchtime call to Mary's. One of the crew was not quite finished with the noon meal, so they took it with them on the call. Finishing with his lunchtime treat as we arrived to the scene, the crewmember daintily dabbed his mouth with a blue paper towel and left his plate on the rig. The couple called us because Mary had sustained a very minor laceration to her hand when she broke a plate while doing the dishes. After disinfecting and bandaging Mary's wound, we inspected the couple's plates and dishes and noticed most of them were made out of ceramic or glass; and all of them were chipped. Someone retrieved the dirty plate off our rig, and over the next few years, we brought the couple a few of our plastic station dishes when called to their apartment. It was the same concept many gas stations employed during the blissful time of my childhood when full-service filling stations not only pumped your gas, checked under the hood and cleaned the windshield, but they also gave their customers a place setting with every fill-up. Before long, Tom and Mary had more of our dishes than we did.

Our fire department family has had this kind of relationship with our customer family for the past 30 years, and I don't see anything on the horizon that's going to change it. I don't know of any system that has all the answers (or even some of them) concerning how you provide service to a set of customers whose problems fall outside our traditional "put the fire out" or "save your life" service delivery models. The fire service has been struggling with this dynamic for quite sometime. After our organizations select the best and the brightest new members, we run them through very intense training programs. Most of a new member's training time is spent learning their role in an organized fire attack, emergency medicine, and some of the more basic concepts of fire station and apparatus maintenance. Very little, if any, time is devoted to what has become the majority of our runs. This service has just snuck up on us.

Our service delivery menu has come to look like a mystery crate a person buys at a bankruptcy auction. Only one or two of the contents are clearly identified, but you bid on the whole lot. We went into the EMS business with our eyes wide open. During our early foray into the world of EMS, we all held varying opinions ranging from violent opposition to dressing veteran firefighters in modified nurses uniforms. The 30-year report card shows that EMS has strengthened our position in the community and has given us new leverage at the bargaining table. Many of our departments have been able to stave off cuts because we took on EMS. It's impossible to calculate the number of jobs that have been secured (and not lost) because we got into the EMS business. Without intending it, expanding what we do for the community has made our departments (or family) healthier.

We now operate in a world where there is more competition for the billable pieces of our service (i.e. medical and emergency transportation). The private, for-profit fire department (part of Rural Metro Corporation) that serviced Scottsdale, AZ recently threw in the towel because fire suppression became economically unfeasible. This happened just weeks after they won a victory in a hotly contested election where the community voted in their favor to continue as the lone service provider in lieu of forming a municipal fire department. Ain't it funny how life works out?

One of the major challenges we face is what to do with the people who call us for service who don't fit into the nice, neat, round and square holes we traditionally fill. Most of these customers suffer from chronic conditions: They are chronically mentally ill, chronically poor, chronically addicted or just plain chronically maladapted in a way where they cannot navigate the complexities of today's society. This isn't going to be the next big chunk in our service delivery package; it's our current 400-lb. gorilla.

Caring for the "chronic" will end up being a very large part of the multi-gazillion dollar healthcare system. My home state of Arizona pays tens of millions of dollars in contracts with private providers. Because we are an essential link in this system, the prospect for funding exists. I think there will be more opportunity in developing the systems our fire departments use to manage the delivery of this service than there currently exists in chasing federal homeland security grants. It seems the cops and the beltway bandits have gained a chokehold on those dollars.

When we look back on why we are successful, it all flows from how well we treat and take care of the customer. Everything else is secondary. Our pay scale, benefits, leave package and pensions are directly connected with the service we provide to the community. Our organizational health will continue to be tied to our quick response and ability to service the broad range of customers who make up our communities. Dysfunction has never been so much fun or held so much promise.

Chapter 10
Hanging in the 'Hood

Last month I wrote about family-related issues. To quickly review: When a person joins a fire department, they become a member of their third and fourth families. The first two families are the same that most other humanoids belong to the one you are born into and the one you create through holy- or state-sanctioned matrimony. The third family are your fellow fire department members. The fourth family is the fire department customer base. In most cases, they are more like a set of distant cousins you see at weddings and the occasional family reunion. In this article, I'd to take a closer look at the issues firefighters face when navigating the minefield of belonging to a traditional "nuclear" family and a fire station family.

Most training-academy graduations are attended by mothers, fathers, brothers, sisters, grandparents, aunts and uncles. Boyfriends, girlfriends, husbands and wives often join this group of blood relatives. Every member of the audience has one thing in common—they beam with pride over the fact that their loved one has become a firefighter. It is a very noble profession. This is one of the happiest days a parent will ever have because they know their child is embarking on a lifelong career of providing an essential service to the community. The joy becomes overwhelming upon the realization that little Johnny or little Suzie is all grown up and will receive an ever-increasing paycheck every two weeks for the rest of their working life. Mom and Dad can now turn their child's old bedroom into a home office, game room or love shack.

Each firefighter has a unique multi-family dynamic. On one end of the spectrum, you have very young firefighters who are just beginning their life's work. Many of these young members begin their career as single people; at this point in their evolution, life is all about them. Just the other day, a 20-year-old firefighter asked me how much responsibility went into owning a dog. After a brief conversation about his lifestyle, I recommended he buy a fish and flush it down the toilet as soon as he got it home from the pet store. This would eliminate the hassle and expense of purchasing an aquarium, accessories and fish food, and the emotional baggage that goes along with slowly killing an aquatic animal over a period of time. People who only spend a couple days a month at home shouldn't own pets.

Carefree doesn't even begin to scratch the surface of what it feels like to be a young, single firefighter. Most of us work a shift schedule where everyday we work is like a Friday followed by a two-day weekend. After a while you lose track of the days of the week and end up standing in front of the bank wondering why it's closed only to figure out it's Sunday. This will change for all of you young whippersnappers when you have school-aged children. Young, single firefighters tend to hang with their workmates on their days off. The combination of spending 24s together at work and then large amounts of off-duty time together further deepens the family bond.

Many firefighters begin their careers with family connections already in place. I think the largest group on my department sharing common DNA is nine. Up until a few years ago, five members of this family worked at the same station on the same shift. It was rumored that their parents had taken up residence in the turnout shed behind the station. Coming from a fire service bloodline means everyone in your ancestral family understands the firefighter lifestyle. I saw a poster the other day that pictured a pride of lions laying in the savannah. The caption read, "We promote family values almost as much as we promote family members." The fire service has a long history of being a generational occupation. A child following in their parent's footsteps has the advantage of common experience. Much of fire fighting's mystery falls away for offspring who have been baptized with tank water. A parent joins the fire service for the second time when their child follows their career path. Some of our members join our ranks with a bucket full of experience from outside the fire service. Our departments hire cowboys, cops, electricians, stockbrokers, schoolteachers, and circus clowns as entry-level firefighters.

I have a friend who has been on the fire department for six or seven years, this being his second career track. Prior to joining the fire department, he spent more than a decade as a very successful and highly regarded divorce attorney. Lawyer jokes aside, imagine spending your working life around people who dislike one another so much they are willing to go through the process of legally dissolving their relationship? I'd rather go on a nonstop barrage of difficulty breathings and ill diabetics.

My ex-lawyer acquaintance made his career change

after one of his parents became terminally ill. During this sad period, the fire department responded on his ill family member on multiple occasions. After this life-changing event came to a close, he did what most clear thinking, intelligent people do—he reflected. One of his most profound recollections was how professional and nice the firefighters had been while his family went through a miserable time. He compared that against working with grown people fighting over who got the house, car, dishes and the credit-card debt. Even though he was the other side of 40, out of shape and had a financially lucrative career, the decision came easily. He got in shape, learned everything he could about the fire department, tested for the job and won. Now he has it all: a career that fills him up and a built-in workforce (or family) that keeps him as busy as he wants to be at his divorce-lawyer side job.

A career in firefighting shapes your outlook and perspective. Although a firefighter's life has the same basic ingredients as anyone else's, firefighters quickly discover that in life, there are no guarantees—especially when it comes to the really important things. There is a big difference between a bad day that results from someone screwing up the paperwork, the drill press breaking down for a couple of hours, the sale not happening and one that happens as a consequence of pulling three dead people out of a second floor apartment, having two kids drown in the same green pool, or finding someone's severed arm in the median of the roadway. Our work glows with a different perspective. This leads to situations where you see events in a different light than your non-firefighter loved ones. The following incident speaks for itself.

Ten years ago I was assigned to my fire department's video unit. Our charge was to produce an hour-long weekly TV show. This program was broadcast on a closed-circuit cable channel. The show featured incident critiques, the news (an overview of last week's incidents), training topics, exciting interviews and human-interest pieces. During this phase of my life I lived in a very nice middle-class neighborhood with my wife and three kids. All the neighbors on our street belonged to the same census tract. We all had young children, promising careers and brand new houses. Throughout the years, we all got to know one another. All the other husbands had regular 9-to-5 jobs. I was the anomaly, working odd hours and getting to wear T-shirts to work. Despite explaining my schedule dozens of times, I don't think any of them ever understood my work hours let alone the nature of my work. This all changed for my neighbors one sunny spring afternoon.

I was at work, sitting in front of hundreds of pounds of computerized video editing equipment, slowly baking my brain with electronic energy, when the phone rang. I answered the phone, and my wife greeted me. During the course of our relationship, we have had a variety of phone conversations. Most of these have been very pleasant exchanges. This was the first call in 10 years where my wife started off in high emotional gear. My first thought was some tragedy had befallen one our children. After I asked my wife a few times about the state of our kids, she finally blurted out, "Our kids are fine, I think our neighbor Bill is dead. Would you please come home?"

I had to restrain myself from laughing because for the last 30 seconds I thought one of my young children was either maimed or dead and now my wife had just confirmed that neither of those things was true. In fact, she told me, "Our kids were fine." The situation in my mind leapt from the death of a loved one to the tragic end of a neighbor whom I barely knew. I told my wife I would be home shortly and hung up the phone.

I don't think my neighbors would have understood my reaction. None of them had my perspective. They had never been members of a fire-department family. This is not a value judgment, just a fact. The residents of my street were decent, hardworking people who sold insurance, computers and houses, or owned air conditioning companies. None of them possessed a career where they worked around buildings exploding in fire or dealt with people who were dead, maimed or somewhere in between. As I drove home I pondered my reaction—was I twisted, or were the residents of my street hypersensitive?

As I wrestled with this issue, I stared into the sun and had a flashback to a story a friend of mine shared with me. This individual was a captain-paramedic assigned to an engine company. His station was located within a mile of his house, so he had a short commute to work every third day (I'm sure his schedule confused the neighbors on his street, too). One day his engine was dispatched to an auto accident. As his engine company arrived at the intersection, he saw a badly damaged vehicle that appeared to be his wife's car. In an instant, he realized that his wife and two daughters would be driving this very route on their way to one of the child-related activities scheduled for that day. He came off the rig in a near panic, convinced he would find his family mortally wounded as he rushed toward certain heartbreak. He didn't feel the slightest bit odd that his life's greatest relief and joy came from looking at two total strangers who happened to be as dead as doornails.

As I rounded the corner toward home there was a beehive of activity. I saw my wife Michele with a group of neighbors and went over to find out what had happened. Michele broke away from the group long enough to tell me that our neighbor, Bill, had committed suicide. Bill was the most popular guy in the 'hood. He was a very successful professional—young, married and a father to a new child. The American dream. As I headed toward my house, I ran into our department's chaplain. He filled me in on the afternoon's events. "Your neighbor hanged himself in his garage."

Our community mailbox was directly across the street from Bill's house, and every afternoon most of the neighbors would get their mail and shoot the breeze. Half a dozen residents were gathered around the mailbox when Bill's wife drove down the street and turned into her driveway. I'm sure she smiled and waved at the congregation assembled around the mail drop just before completing the turn into her drive and pushing the button on the garage-door remote. The crowd across the street was surprised to hear her shriek and tentatively headed that way. Later on my wife told me they saw Bill standing in the middle of his garage with his head cocked to the side and thought he was pulling one of his well-known pranks on his unsuspecting neighbors.

The group of mail-retrievers made their way to Bill's wife and attempted to calm her with their "Bill's being a prankster" explanation. While part of the group tried to comfort Mrs. Bill, several of their counterparts ventured into the garage to call Bill on his bluff and scold him for upsetting his wife. They came face to face with a human being who had hanged himself hours earlier. I watched the neighbors' varied reactions as the chaplin gave me the details of what my wife and her friends had witnessed. Bill had tied a noose made out of a nylon tow cable to his chin-up bar and around his neck, and then he stood on his lawnmower. After he pushed it out of the way, his feet dangled only a few inches off the ground. He hung there long enough that his neck stretched and his feet settled to the ground. His head was lying on his shoulder, and all of his facial features were blown out of proportion and discolored. It looked like a criminally insane cartoonist had chewed off his own hand and painted his idea of what a hanged person would like. It wasn't a pleasant sight.

The two of us stood there for a few minutes, listening to and watching the neighborhood emotionally unravel while the cops finished up their investigation. After Bill's funeral, the chaplain and a firefighter grief counselor came back and met with the neighbors. Everyone said it helped, but the old 'hood never really recovered from their day with life's ugly flip side.

Compare this to an incident that happened a few months later with my other family at Phoenix Fire Station 11. It was around 10 p.m. and our ladder tender had just headed west on a call. My brother John was driving and was a couple hundred feet down the road when he noticed a vehicle driving erratically in the eastbound lane. As the car approached, it picked up speed and crossed four lanes of traffic prior to hitting the ladder tender head on at a high rate of speed. John saw that the car intended to plow into his apparatus and had managed to get his rig stopped prior to impact. The crew momentarily sat in stunned disbelief and made sure that everyone was all right before the captain radioed in for a 2-1 medical. The rest of the crew climbed off the rig and went about treating and extricating the driver of the vehicle that had just hit them.

People who drive into large, heavy objects at high rates of speed do not have a very high survival rate, and this was no exception. Despite running into a very capable extrication crew, having top flight paramedics show up seconds later and choosing to stage his accident directly in front of a very busy trauma center, the driver of the vehicle died. Later on that evening, the police determined that the driver of the vehicle had committed suicide. The remainder of Ladder 11's shortened shift was spent talking to police accident investigators and the usual array of fire department supervisors and emotional support staff.

The next shift, our shift commander stopped by the station to see how the crew was doing. We were all sitting around the table, following the morning ritual of coffee and fellowship when the highest-ranking B-shifter entered the room and asked how everyone was holding up. One of the engineers who had been on the ladder tender the shift before responded, "Chief we are having an excellent day." He then patted my brother on the back and said, "We've already been on one call today, and John hasn't killed anyone."

Chapter 11
Electronic Brain Nail

Most things come with an upside and a downside, and the electronic revolution is no different. In the span of two decades, we've developed cell phones, the Internet, e-mail, electronically controlled discharge valves, PDAs, CAD systems, MCTs in our response vehicles and cordless hair removers. These devices free up time, allow us to access information more quickly and provide painless, total body hair removal for many A-shifters.

Cell phones and the Internet have completely changed the world. Consider the hotel industry, which has lost millions of dollars because their customers opt to use cell phones in lieu of in-room telephones. For years, a 10-minute phone call made to any location outside the hotel cost more than the room. Soon many travelers figured out they could eliminate these exorbitant phone charges through the use of prepaid calling cards. Just like everything else in life, this cost-savings tool has a downside. I remember sitting in a Holiday Inn watching my father punch a set of numbers into the telephone. The number-pushing process took so long that the veins in his forearms bulged and I passed out on the stained, blue carpeting from the sheer exhaustion of watching him. Then cell phones came along and completely killed the hotel industry's long-distance telephone scam.

The earliest cell phones looked more like sewing machines than today's sleek versions, and oddly enough, they were instant status symbols. Technology evolved, and soon people were walking around with ear buds the size of small tumors attached directly to their ears via a dubious black wire. There are few sights more obnoxious than a person clutching the mouthpiece on their ear-bud wire as if it were a set of rosary beads while they mindlessly stammer away, oblivious to anything happening around them. But at least these visible earbud wires allowed you to identify the users as cell-phone addicts. Blue tooth earpieces are much harder to detect, so it appears that these people are simply talking to themselves, which has always been a sign and symptom of mental instability.

My department put mobile phones on all of our apparatus more than 15 years ago. At that same time, some fire department staff received individual cell phones as well. We added the phones to our toolbox to increase our communications capability. In the early days, they came in handy when a customer needed to tell their family they would be a little late because they had just been in a car accident, or when a homeowner wanted to call their insurance agent after a house fire. They were also helpful when we needed to have a long conversation with the alarm room, another unit or a battalion chief. They allowed us to say things we might hesitate to say over a tactical radio channel and eliminated the need to pull off the road to use a pay phone. These mobile phones also prevented an undetermined number of fire station kitchen fires. Who among us has not called the in-quarters or next-due company to have them turn off the oven when we were tied up on a call?

During this bygone, pre-caller ID era, it wasn't

uncommon to see fire-department schmoozers and rectum weevils make personal calls on their city-issued cell phone while a perfectly good landline sat within arm's reach. (The younger generation will have to pardon my rant. This was back when each cell-phone conversation cost more than the price of delivering a package.) Employer-provided cell phones belong to the employer. For fire department employees with fire department cell phones, this means every call they make or receive on their cell phone is public record. Each month, the department cell-phone bill lists all the people and places you communicated with that month.

This record is available to anyone who wants it: your boss, your spouse, the law and the media. And yet we're surprised when the media obtains the department's phone records under the Freedom of Information Act and discovers some minuscule percentage of city-funded cell-phone calls fall into the "questionable" category. This is a simple intelligence test for members who make mobile calls on the city's dime. All it takes is one simpleton making a few calls to 900 numbers, Madame Shirley's House of Massage or their Eastern European bookie and spiritual advisor to score the local investigative reporter the top slot on all five weekly news casts.

My employer solved this problem by replacing staff and middle manager (chief) cell phones with a monthly communications allowance. This made a lot of sense since most human beings old enough to talk already own a cell phone. This system allows the users to talk to whoever trips their trigger without the hassle of the city tracking and being responsible for the tens of thousands of calls made every month.

I didn't start this essay off intending to complain about the overuse of cell phones, but in today's world many stupid human acts include cell-phone use. Last year my department responded to a traffic accident where the driver of a regular pickup truck rear-ended a semi carrying a load of steel. The only serious damage to the pickup truck was a large hole through the windshield, the product of a large I-beam. The same I-beam delivered even more spectacular damage to the pick-up driver's head. Our crews arrived to the scene, determined that the driver of the pickup truck was forever dead, provided short-term comfort to the distraught semi driver and waited for the folks from the morgue to collect the dead body. The rest of this story would fall under "urban legend" category if I hadn't heard it firsthand from the on-scene battalion chief, who got it directly from the driver of the morgue van.

The dead driver was placed into a body bag, loaded into the back of the morgue van and was en route to the coroner's office when the driver and attendant heard the muffled ringing of a phone. Both of them checked to make sure it wasn't their cell phones. Their attention shifted to the dead guy wearing a body bag in the back of their van. I hold a deeply personal belief that one should never answer a dead stranger's ringing phone. The morgue workers didn't subscribe to this philosophy and pulled over to secure the ringing device. Even though the ringing stopped, they continued their quest and went into the bag, determined to retrieve the phone. The crew searched the dead man's pockets and found nothing. The phone began ringing again. The sound was coming from the corpse's head. This explained what caused the fatal accident.

The major problem with cell phones is they cause us to divert our attention from what is right in front of us. In the case of the pickup driver, it caused him to meet a premature and grizzly ending. This is an extreme consequence; in most cases, cell-phone caused diversions are not fatal, just rude. Don Abbott, our training Pooh-Bah, has done several communications studies at our Command Training Center analyzing the radio communications that took place during fire simulations. In most of the simulations, half of the stuff said over the radio was categorized as nebulous and unnecessary. I believe if the same study were applied to all cell-phone conversations, the amount of mindless chatter would be substantially higher.

The Internet mirrors cell-phone usage in many ways, especially the opportunity to use it for stuff considered outside the realm of official business. The computer revolution has created an almost infinite zoo where you can find spiritual enlightenment, watch coeds cook dinner and take showers, research any topic in human history, get the lowest price on a plasma TV, gamble away your life savings, find a soul mate or trace family genealogy back to Stonehenge.

The Internet provides the most complete and comprehensive example of the global community. It is the only place I know where man's greatest works of art, literature, religion and historical record swim side by side with shopping, online banking, chat rooms and blogs, cheap drugs from Mexico, advice on when to trim back your roses if you live in a desert climate, pornography and swift access to most bureaucracies.

The Internet has everything to offer, and for the most part it's completely unregulated. The most fascinating information I've discovered on the World Wide Web is quite shocking. Both of the George Bushes may appear as Homo sapiens to the casual, uneducated eye, so you can imagine my shock when I read online that they are not human beings after all, but members of an underground race of giant lizards. These lizards control all human activity with the electronic energy produced by our televisions, cell phones, computers and electric toothbrushes. I believe the human spirit to be indomitable, and for this reason it can't be long before someone captures and unmasks one of these giant lizards and shows the world their naked writhing body via the Internet. I find it difficult to believe that I had to find this information on my home computer because the City of Phoenix blocks these sites (and many, many others) on my work computer. Could this all be part of the giant lizard plot?

Just like a cell phone bill, all Internet activity is recorded somewhere. As government employees, someone can review anywhere we've navigated on the World Wide Web while sitting at our workstation. Much of the Internet content falls outside the government mantra of

"For Official Use Only." I do not have any studies to back me up on this, but when given the choice, many of our members would opt to look at the nude human form over any other available image. Many municipalities and businesses have taken a preventative approach to this problem. Most publicly funded computer systems have some type of "net nanny" standing guard. Anytime you try to visit a Web site that contains one of the forbidden words or phrases, the system sends you a curt message stating you can't go there.

Cell phones and the Internet are a lot like prostitution. Everyone talks about putting the painted ladies of the evening out of business, but it never happens. As firefighters, we often cross paths with prostitutes. They call on us for help when they have some type of medical emergency or their home or place of business catches on fire. We respond to their emergency, do our job and treat them nicely. The community expects us to deliver service to everyone in the community, but they have serious reservations about prostitutes delivering service to the members of their fire department (at least while we are on duty). These ramblings are in no way meant as a stern moral dialogue. This is a free country, and what a person does in their free time is their own business. We all have this right—without the media reporting on our personal business. But what we do while we are on duty is another matter altogether. I have yet to see an instance where the community reacted positively when a firefighter racked up big cell-phone bills while conducting personal business or spent his downtime checking out Dutch Girl on the Farm Web sites.

I grew up in the fire service listening to my elders preach about the "headline test." This is a simple concept: Plug any action you are considering into the headline of tomorrow's newspaper. If you can't defend it, don't do it. Within the last year, a high-ranking law enforcement official in one of my neighboring cities was forced to resign. His downfall was the close working relationship he had developed with a co-worker. The story ran its predictable course. He denied the charges, the city put him on administrative leave, and the investigation was on. The final act played out on the evening news. The smoking gun was a set of room-service bills from a city-paid trip that he and the co-worker took together (the news crew got a copy of the bills to show on TV). The bills were for a set of early-morning breakfasts and a pair of late-night desserts. Her bill, his signature.

He flunked the headline test.

Chapter 12
Nothing but Static

I recently attended one of the big fire-service conferences. During a break in one of the workshops, I chatted with a firefighter whose department was part of a regional response area that was replacing their computer-aided dispatch (CAD) system. I don't remember the exact cost he quoted, but it was in the neighborhood of 20 million U.S. dollars. He ended his story with the hope that the new CAD system will work; the last one never went online. But on the bright side it cost less than $12 million. Isn't that crazy? This bears repeating: $12 million dollars for a computer system that never worked, not even for one single day. Our country spent millions of dollars to put Martha Stewart behind bars for perceived improprieties resulting from less than $300,000 worth of stock transactions while large electronics companies continue to peddle and charge exorbitant amounts for communications systems that never work right.

My department is experiencing this firsthand with our new state-of-the-art 800MHz radio system. We have yet to go online with it, and word is it will end up becoming some type of hybrid 700MHz system. (Or was it 900?) One of the departments in our automatic aid system has been using their 800MHz radio system for almost a year now. In addition to carrying a radio, they must also carry a pager and a cell phone to communicate in dead areas without radio coverage. Our communications technology has evolved (or devolved) to the point where we can use our portable radios to talk to a space-shuttle astronaut, but we can't talk to a firefighter in the next room. What do you want for tens of millions of dollars?

Our current technology situation has many layers, but I believe there are three main pieces to this puzzle. No. 1 (and most important): Our service really doesn't understand technology. Electrical devices continue to mystify us. Even the most highly qualified paramedic could not tell you what makes an automated external defibrillator work. I'm old enough to remember fondly the days when we kept the heart monitor/defib device in a large, orange plastic box. The top of the box had a secret compartment that stored a telephone. Several large automobile batteries, connected with miles of brightly colored wires, powered the device. It was so heavy it required two paramedics (or a single truck company firefighter) to carry it. I recall old engineers starring in amazement as young, pretty paramedics adorned with gold jewelry and $50 haircuts shocked heart-attack patients back into normal sinus rhythm as their partner talked into the magic phone. Whenever anyone asked the medics whom they were speaking to, they would merely look down their noses at them and say, "God." I suppose this makes them all a bunch of little Moseses. I'm drifting again.

Most of us don't have the slightest idea what the following things are: a megahertz, the level of RF attenuation, a non-repeating trunk repeater, a digital pack, analog, UHF, VHF, simplex, bi-directional amplifiers, uplinks and what a hen weighs. I gained the majority of my electronic knowledge from Hollywood movies.

Much of my generation has been taught that time travel is possible given you have a big enough budget and the future governor of California. Our current reliance on all things electrical is reinforced by the operability of many electronic devices and systems we use on a daily basis. We naively think that when we pick up a fully charged portable radio, key the mic and speak into the device, someone on the other end will hear us. We believe this because our previous radio system worked that way. Our old radio system never honked or bonked when we depressed the push-to talk-button. We could live with situations where the folks a couple Zip codes away couldn't hear us, but the company one room over, the truck company on the roof and the incident commander in the street could talk to one another.

This is the minimum radio system requirement for any and all hazard-zone operations, which brings us to point No. 2: It seems that some people who design, build, install and sell us these high-tech radio systems do not understand our minimum radio-system requirements. It took tens of millions of dollars, a couple of years and several meetings with our radio vendor for them to learn that the partial radio coverage provided by a trunked radio system was not going to be the centerpiece of any communications system we use to manage active hazard zones and firefighter safety.

One has to wonder how a company whose business it is to build communication networks for the fire service could be so clueless about how we use and depend on our radio system. Our relationship with radio vendors is based solely on radio systems. It's pretty tough when we don't understand what they do, and they don't understand what we do. Dr. Phil would call this a dysfunctional relationship.

As if the first two issues weren't bad enough, things get further complicated with point No. 3—system scale-up. This happens when we learn way late and deep into the system's construction and testing phases that our new high-tech, digitally trunked, non-simplex, 800MHz system has a minor problem with the "Can you hear me now?" part because we can't communicate via radio in certain areas of town and inside many of the buildings we operate in. This is one of the inherent drawbacks of trunked radio systems: If your handheld radio can't transmit to the trunk site, you don't get to talk. You won't find this information in most of the radio companies' literature, but don't fret. There is a solution: more trunk sites. Trunk sites don't grow on trees, and most municipalities do not have extra funds lying around to purchase more radio infrastructure for a project that most powers-thatbe think cost way too much to begin with.

These new trunk sites will probably have very little to no effect in making your radios work inside certain buildings. To correct this problem, you have to identify the buildings your portable radios cannot transmit from and retrofit those structures with still more expensive electronic radio booster devices. None of these devices will have a manufacturer's warranty in the event the building ever catches on fire because a building's built-in electronic devices are always among the first things to fail in a fire. The only exception to this rule is very loud, obnoxious fire alarms, which have the opposite problem. No matter how hard you hit them with an axe or sledgehammer, they will continue to make more noise than all of Mexico on Cinco de Mayo.

These electronic tales are all too common, and I can't offer any plausible explanation of how they keep happening. The Phoenix Fire Department's saving grace came from a very talented pair of our own members. We were very fortunate that Mike Worrell and Andy Macfarlane became an integral part of the 800MHz radio project before our fire department went online with the new system (almost two years prior to our switch-over date). They determined the system had inherent shortcomings our fire department couldn't live with and were able to work with the city's IT department and the vendor to formulate a solution that will provide our fire department reliable radio communications into the future. We opted to forgo spending tens of millions of dollars more on a system that couldn't be guaranteed to work as well as our current system, so we've decided to stick with what we have (a simplex VHF system), continue to support it and see what the future brings. Our police department will continue using our city's new 800MHz trunked radio system. The new system offers our police department much more capacity, which was one of the major reasons our city went out and got a new radio system.

Mike and Andy are very unique because they understand the details of our radio needs and the technical details of radio systems. Since the first draft of this article, the NFPA has come out with a Standard 1221 that recommends trunked radio systems should not be used for fire department hazard-zone operations. Mike and Andy continue to be technical visionaries.

Can you imagine if the people who brought us the trunked 800MHz radio system had built the Skynet computer system that put the human race on the brink of extinction in the "Terminator" movies? If Skynet had been built around a trunked 800MHz platform, the world's population would have tripled, Sahara Conner would have had 372 descendants and the Terminator cyborg robot machines would have been sent into the past to destroy a certain communications company. (Author's note—I am not interested in causing turmoil in my editor's life, so plug in any communication company name you see fit.) This brings this article to a fork in the road...

California's governor, Arnold "The Terminator 1, 2, & 3" Schwarzenegger, has caused quite a commotion with his desire to change the state's pension system. California, like every other state, doesn't have enough money to fully fund all of the various state programs. If the state could make smaller pension payments for their teachers, firefighters, police officers and all other government employees, they would have more money available to spend on other things. Try this concept for yourself. Send your mortgage company half of your house payment every month. Spend the other half of the money as you and your family see fit. About 10 minutes after the bank owning the mortgage on your house receives your adjusted payment, they will call you, wanting to know why your house payment is light this month. Tell them that you have adopted a new economic plan that will strengthen and secure our country's future. You are investing half of your regular house payment into the economic engine that drives the good old U.S. of A.

Assure them this will be good for their financial institution, and you have the spreadsheets, pie charts, projections and brand-new Cadillac to prove your point. I predict that in six months, you will be homeless. Along this same line of thought, the current governor of the great state of California shouldn't forget what happened to the last governor of California. While I am on the subject of pensions, I would like to spend our remaining time discussing the president's Social Security plan. Talk of allowing people to have their own retirement accounts under the Social Security umbrella has sparked a vigorous national debate. I read an editorial in my local paper that represents one side of this argument. The writer felt that young people should quit referring to the money they pay into the Social Security system as "their" money. This individual went on to explain that the current payments going into the fund are her money, and the only change needed was for young people to get over it and move on. Sounds like a classic pyramid scheme to me. The argument on the other side of the hall is best explained by a recent poll where a large number of young people felt they were more likely to see space aliens than future Social Security payments. Let's not get lost in the debate; instead we should focus on comparing what a person gets for their money. Most firefighters have some type of pension plan they pay into; neither they nor their employer has to make payments into the Social Security system. Pension plans vary from department to department, so I will use my department's pension system because it is the one I am most fond of and familiar with. One of my regular paycheck deductions is the 7.65% pension contribution. I have been paying into this pension fund since 1980, depositing enough principle into the fund to pay the average salary of a Major League baseball pitcher for at least a couple of games. Each payday, my employer (the City of Phoenix) also makes a contribution. This amount currently hovers near 7%. This is close to the same contribution we would both be making if I were in the Social Security program instead of a public safety pension system. For all practical purposes, the only thing my pension has in common with Social Security is that the employer-employee contribution rate is pretty similar (within a percent or two). Here are some of the major differences with my pension plan:

• After 20 years, I can retire at 50 percent of my yearly gross pay (an average of your best consecutive 36 months).

• My monthly pension payment increases each year I stay beyond 20 years, up to a maximum of 80 percent at 32 years.

• Our pension system permits its members the opportunity to participate in a DROP program. A DROP program basically allows a member to set up an additional retirement account at the end of their career.

It would require a full article to explain the details of the various DROP programs. DROP programs are good for the employee, employer and the pension fund. I don't anticipate anything resembling a DROP program resulting from the current Social Security debate.

• My current pension system is actuarially sound. Prior to the stock market's wrong turn a few years ago, it was127% funded; it is currently around 95% funded.

Some of the Social Security debate includes the argument that police officers and firefighters have it too good; our pensions are out of line and are costing the taxpayers too much money. This is not true. If we didn't have our own pension systems, we would end up paying approximately the same amount of money into the Social Security system. This logic winds itself down a twisted, nasty path. It's as if they're saying, "Social security is not nearly as good as your pension. We think it is time to make your pension as bad as Social Security." Wouldn't it be smarter to make Social Security as good as our pensions? Every argument you hear about Social Security goes all over the page. Some say it will be bankrupt in a few years, while others claim it will be stable into the next century. The entire debate is ripe with the stench of partisan politics, and it doesn't appear that anybody is interested in positive change so much as they are in gaining political mileage. A few months ago I suggested that we send the fire service's Kitty Men to Washington to fix our budget mess. I think they should take Mike and Andy with them to begin fixing the Social Security system and the current radio debacle.

Chapter 13
Ying & Yang of Competence

I am assigned to Fire Station 30. As of a few years ago, Station 30 sat in the geographic center of Phoenix. This station houses Engine 730 (an ALS adaptive response unit), Engine 30 (a BLS engine company), Rescue 30 (a BLS ambulance), North Shift Command and our department's backup command van. The neighborhood is best described as eclectic. It has a varied mixed of socioeconomic and ethnic groups. Today I am working an extra shift with the C group. They, too, are best described as eclectic. Over lunch, one of the young C-shifters was lamenting that they hadn't had a fire in a while. His captain teased him: "You don't need to be worrying yourself about your lack of fire. You are here to take care of Mrs. Smith." The young paramedic/firefighter smiled at his captain and said, "I got some bad news for you Skip, Mrs. Smith is dead. She died because you and your generation killed her, and despite me and my homies' best efforts, it doesn't look like we can bring her back. You have angered the fire gods and now I must pay for your sins." We all laughed hard and went back to our tasty BLTs.

After lunch, the group went behind the station and worked with their probationary firefighters on basic task-level skills. As the tribe threw ladders and laid hose, I stared into the sun and reflected on the fact that we devote the majority of our training time to the things we do the least. When one considers the high hazards we face in our work, this is just the price of doing business. As I watched the group do its thing, I contemplated the question of what constitutes firefighter competence. When I began my career, if there was a single requirement for being a good firefighter, it was your ability to press a fire attack. If you couldn't do this, the group considered you worthless. An individual could be flawed in every deplorable, disgusting way imaginable, but if they could advance an attack line into a burning building, they were considered competent. If that same person knew the tactical nuances of how to quickly get through locked barriers, the proper attack points and positions, when to stand and when to duck, and all of the other ingredients that go into a well-executed fire attack, they were considered outstanding.

I don't think this basic requirement has changed in the last two and half decades, but the list of qualities that goes into being a good firefighter has expanded. As long as we have a fleet of vehicles designed to flow water and our uniforms say "Fire" on the back of them, the ability to perform on the fireground will remain a job requirement. As with any other occupation, some of our members will be more skilled at certain aspects of the job than others. We shouldn't use the performance level of our most outstanding firefighters to establish the group's benchmark. This is akin to the NBA expecting all of its players to perform like Michael Jordan. If this were the case, there wouldn't be enough players to field a single team.

Standards are lovely things and provide us parameters for evaluating a whole host of job responsibilities, but they are oftentimes very inaccurate, as the following story demonstrates: During a simpler time, my fire department conducted all of our departmental training out of our training academy. Companies went out of service first thing in the morning and drove their rig to the training academy, where they spent the day doing that quarter's training. The instructors were typically our department's resident experts for that given training block. If we were doing strategy-and-tactics training, a battalion chief or shift commander would instruct. Hazardous materials was taught by one of our hazmat technicians, rope-and-knot classes featured one of our granolaeating, high-angle rescue gurus and pre-eclampsia childbirth lessons were delivered by young, smooth-skinned paramedics. You get the idea. Most of us did not burst with anticipation over our quarterly trek to the training sessions, but like most fire-department activities, some part of the day's event was going to be memorable. (As proof, today I'm writing about something that occurred 20 years ago at one of these gatherings.)

The training topic de jour was EMS, and the first instructor out of the gate was a captain/paramedic who was well known for his obsessive-compulsive disorder. He conducted his life much like the Felix Unger character on the "Odd Couple" movie and popular TV show. He was (and still is) a highly qualified and very competent paramedic. He was (and still is) a very anal retentive C-shifter. If someone in my family were having a medical emergency, I'd be happy if he was a member of the responding crew. That doesn't mean he's a good instructor or that I would like to spend a neurosis-filled 24-hour shift with him.

Captain Felix Unger began our class with a stern lecture about germs, bacteria, viruses and the inappropriate washing methods B-Shifters use to clean their hands. He continued talking down to us mouth-breathers during a lecture where he reviewed the application of the "Thumper" on a Resusci Annie life-like torso doll (sans most lifelike human orifices). For those of you too young to know what a Thumper is, it was a fascinating piece of emergency medical gear possessing the streamlined design characteristics of a Russian dump truck. It was made up of half a dozen components that never seemed to fit together very well, and adding a dead person to the equation only made matters worse. The centerpiece of the contraption was a piston-driven CPR hammer and a jet-powered ventilator that hooked up to an intubation tube a paramedic would snake down the dead person's throat. The device—powered by large 02 cylinders— was assembled on a dead person right after we injected enough liquid baking soda into their lifeless body to make biscuits for all of Florida and part of Georgia.

I've worked dozens of codes where we applied the Thumper and saw only a single instance of short-lived success. Right after we installed the Thumper device onto this particular dead person, they abruptly came to life, tore the intubation tube from their throat (the paramedic at the time called this procedure an "extubation"), looked with large eyes at the plunging piston, then vomited all over our captain before permanently dying. Back to class.

Captain Felix was a big believer in the concept of "training in context" and had removed Resusci Annie's shirt prior to placing the Thumper on her. After he went through the 429 steps involved in Thumper application and had the machine pumping away on the lifeless doll with nasty fake hair, he told the class, "Before you transport Granny out of the house and into the back of the ambulance, show the proper respect and tape a 4X4 gauze pack over each of her nipples so the media cannot videotape her bare breasts and upset the grieving family." Nothing would give me more pleasure than to share the ensuing conversations that took place between the B shift student body and our C-shift instructor, but the people ultimately responsible for the content of this magazine would surely edit it out, which would cause me great consternation. Let's just say the spirited discussion drove our instructor temporarily insane and resulted in an hour long break before our next session.

The day's next instructor was a bigger surprise than Captain Felix but every bit as delightful. As I mentioned earlier, the instructors who taught at our training academy were our department's resident experts on the subjects they presented. Because our next class was on EMS documentation, we expected the instructor would be a paramedic or captain well known for their documentation skills. None of us could have been more surprised to see a captain who was infamous for his inefficiencies when it came to recording the details of an EMS incident. When he entered the room, most of us thought he had merely shown up a little late for training (as a student). We all giggled when he made his way to the front of the room, informing us that he would be that day's instructor. He was the polar opposite of Captain Felix and came from the rapidly fading era of "plain old guts."

It would be inaccurate to say that instructor No. 2 was the Oscar Madison to instructor No. 1. He shared more characteristics with Earnest T. Bass from the Andy Griffith show, and he was our department's poster child for members not fond of incidents of a medical nature. If our training session had been for some type of hands-on firefighting activity, Captain Bass would have been a competent and appropriate instructor. In the event that I need to be saved from a burning building, Captain Bass would be a welcome addition to the team, but he would not be on anyone's short list to deliver medical service to a loved one. Before Captain Bass could get the class started and delve into the nuances of technical documentation for the sick and injured, one of the students asked the question on everyone's mind: "Why the hell are you teaching this class?" The answer goes right to the heart of how human ingenuity can derail many of the systems we use to measure performance.

My fire department's EMS form has four pages, the bottom three being carbon copies of the first page. In the upper portion of the form, we record the date, incident number, assigned units, incident address and the nature of the call. As you work your way down the form, there's a space for patient info, medical findings and treatment—pretty standard stuff. The top copy of the form gets routed through one door of our EMS section and becomes a permanent record of the incident. Two of the carbon copies go to the hospital: One becomes part of the patient's medical record and the other goes to the hospital's billing department. The fourth page of the form goes through another door of our EMS section and is used for quality-control review.

This fourth form (the goldenrod copy) is responsible for Captain Bass's brief appearance as an EMS documentation company-training officer. A group of doctors, nurses and paramedics review page No. 4 and keep a running log on all of the captains and paramedics who fill out the EMS forms. As a captain, I would get back stacks of these forms that I had filled out. Most of them looked like an angry 7th grade English teacher had used a red pen to take out a lifetime full of frustrations and failed relationships ("You misspelled hematoma!"). Much like the confessions mandated by the Catholic church, those monthly dressing downs were not so much intended to make me a better documenter or even a better person, but to make the reviewers feel better about themselves. In retrospect, I hope the entire group of them is stricken with venereal warts.

After Captain Bass's first stack of heavily corrected forms came back, he made the decision to route all future copies of the goldenrod form directly into the trash. As the days turned into weeks and the weeks turned into months, Captain Bass's EMS documentation rating climbed to the point where he was in the top 1 percent of BLS documenters, and he was awarded the coveted Golden Pen of Outstanding EMS documentation. The 1-percenters were further rewarded by getting to teach documentation to the rest of the department (on an overtime basis) during department training.

In true bureaucratic fashion, a non-fire department medical professional selected that quarter's documentation teachers based solely on who had the best overall ratings. This is where the ancient art of redundancy came into play and salvaged future training. The deputy chief in charge of the training academy was a cantankerous old man who knew everyone on sight and didn't take his orders from a group of doctors and nurses and their bell-curve ratings system. He called Captain Bass into his office, congratulated him on his lovely golden pen and summarily fired him from his OT training gig.

This doesn't bring me any closer to answering my initial question regarding what constitutes true competence, so allow me to take a final whack at it. A lot of you may be thinking that I haven't mentioned anything about hiring standards, training, management, supervision or leadership. We all have an overabundance of these things (at least we have lots of meetings where we talk about them for hours on end), yet we still find a wide range of skill levels within our workforce. One of our true strengths is that we operate as companies. Each member brings different talents, experience, abilities, perspectives, flaws, likes, dislikes, idiosyncrasies and personal baggage to the crew. Not all firefighters can jump over tall buildings in a single bound or even get a fully charged 2" line up to the third floor. A little further down the rabbit hole you find firefighters who can do the jumping over tall buildings thing and can drag big, pressurized attack lines up impossible vertical distances, but they can't handle the blood-and-guts part of the job, so they compensate by reading poetry and spooning with their soul mate, Bruce, on their days off. This certainly doesn't describe all of us individually, but since we all hire human beings as firefighters we collectively have a full range of people who make up our departments. We are not an army of one (whatever the hell that means), and our true strength lies in the crew's ability to come together and get the job done.

Synergy is one of the most overused words in the English language, but there is no better phrase to describe competent company operations. When you stand back and watch good companies operate, it seems that they communicate telepathically because no one says very much. Each member of the company goes about doing the business of their own job. All of these separate tasks are choreographed in a way that each individual's efforts somehow fit into a collective solution for the incident problem. This level of performance transcends simply knowing which way the threads tighten, the proper technique for advancing charged attack lines, switching over from tank water to hydrant water, controlling your respiration rate while using an SCBA, along with the millions of other things that go into a simple fire attack.

True harmony comes from knowing the details of your job, understanding how it affects and fits in with everyone else's assigned function in order to make the incident problem go away, then physically performing the right set of actions at the right time. We have all watched companies that always seem to do the right thing. This not only comes from knowing, and doing, the right thing at any given moment, but also setting up the right set of actions (and backup plans) they should take 10 minutes from now. This is very difficult to capture and teach. Years ago, a Supreme Court associate justice trying a pornography case was asked to give a definition of what constituted pornography. His answer was, "I could never succeed in intelligibly doing so...but I know it when I see it." This same sentiment goes a long way in describing the competency levels of all the incident scene players.

Chapter 14
All Hail the Queen

A while back I spent a morning hanging out with a group of recruits. We were talking about the way we manage hazard zones. There was a big portable white board sitting in the corner of the room. It had the group's class number and mascot written at the top. (I never understood the mascot thing.) I flashed back to a decade ago when a recruit class wanted to name themselves the Lucky Sperm because morea than 90 percent of its members were blood relatives of senior department members. The training captains told the group that wasn't going to happen and to name their class something else. I don't remember what they came up with, but I'm sure it was a very manly moniker, something like Dragon Slayers, Flame Eaters, Douche-Bag Mechanics or Necromancer Assassins.

I spent 10 years working on a ladder truck that went by the name of the Raging Queen. This was taken from a very early episode of the cutting-edge television show "Saturday Night Live." The John Belushi-era cast did a skit where they were pirates aboard a ship called the Raging Queen. The motto of their pirate vessel was, "We sail aboard the Raging Queen, where punishment is considered manly" (or something to that effect). They were a merry band of pirating males who loved giving and receiving punishment from one another, along with tanning on the ship's deck and eating banana bread. One of the guys who watched this original episode was a seriously disturbed battalion chief who was a natural artist (in more than one way). During the next shift, he painted "The Raging Queen" on a large placard and then forced a mechanic to attach it to the front of Ladder 11. He held a small ceremony where he christened the truck with the words, "From this day forward, this mighty ladder and her well-groomed and manly crew will forever be known as the Raging Queen." This happened during a time when very few of our department's apparatus had nicknames; most of them were simply known by their number. If you were assigned to the ladder truck stationed at Fire Station 33, you worked on Ladder 33, not Sky Riders (which is a much gayer name than the Raging Queen). Ladder 11's crew took all this in good humor, thinking they would go along with their chief's idiosyncrasy for a week or two then go back to simply being the crew of Ladder 11. They could not have been more wrong. Every single member of the Phoenix Fire Department watched the SNL Raging Queen episode and now made the direct connection between manly pirate punishment and Ladder 11. This pleased the B-shift battalion chief at Station 11 beyond words.

Ten years later, I bid on—and got—the captain spot on Ladder 11. For the next 10 years, I filled my cup with what have become the "good old days" of my career. On many occasions, the Raging Queen placard on the front of our rig pulled me out of despair and saved me from myself. These seething, rage-filled moments of desperation always came after 1 a.m. and were a by-product of delivering emergency service to the lowest rung of the socioeconomic ladder. I vividly recall wanting to administer inappropriate physical action toward members of our customer base—customers I felt were conspiring to drive me crazy with their relentless needs. In more than one instance, I was distracted from this action when the apparatus nameplate caught my eye and reminded me of the bigger picture. I learned it wasn't always easy being the captain of a ladder company named after a fictional gay pirate ship, but nothing worth having ever is. These were the good days, when we cut roofs, overpowered engine crews and took their attack lines, used too much cling-gauze for patients with cranial lacerations—and fought with the C-shifters to keep our rig's christened name unchanged. Not everyone assigned to Ladder 11 thought it was proper to name a rig after an imaginary gay pirate ship.

Several years into my tenure on Ladder 11, our counterparts on the C shift wanted to forever banish the Raging Queen designation. They called everyone they could to try to drum up support. The C-shifters were able to arrange a meeting between their battalion chief, the district commander and our department's apparatus officer. They decided that if all the crews assigned to Ladder 11 agreed, they could change the name to something more pedestrian. The next shift, they pitched their idea to us, stating that A shift really didn't care about the name one way or another, and if we were agreeable, none of us would have to "ride our truck in shame" anymore.

To give this tale the proper backdrop, I should explain that the name-changing debate occurred during a period when the three shifts at our station were having issues regarding the ordering of station supplies. In our supply-ordering system, whenever we were running low on something, we wrote it down on the dayroom blackboard, and at periodic intervals the C shift would fill out the appropriate paperwork, send it through the proper channels and a stock truck would show up at the station and drop off dozens of boxes of paper towels, rig rags, oven cleaner, pens, batteries, etc. C shifts across the city were in charge of ordering supplies. They had the hassle of putting away the dozens of boxes of stuff that showed up every month, but they were also the domestic goddesses who controlled what showed up in those boxes. For example, during this period, the engine and ladder companies housed in Station 27 were being moved to adjacent stations for several months because of a station remodel. During the relocation process, one of the captains needed to go into the attic for some reason. He couldn't believe the riches contained in the station's attic space. The C shifters had built dozens of feet of shelving that held a treasure trove of station items. He found more than 300 boxes of Spic and Span, a dozen kitchen knives, salvage covers, a wide array of attack-line nozzles, dozens of boxes of paper, stacks of coffee cups and crates of road flares—and this was just the tip of the iceberg. It was hoarding on a scale one usually associates with extreme religious cults. This is the power of the ordering clerk. They derive their juice from controlling the flow of the sundry items needed to do the most basic of things. Gaining access to these items can become an exercise in groveling. When the strong need something controlled by the timid and the weak (in this case, B shift needing access to C-shift-controlled items), it can create problems, sometimes to such a degree that the C shifters would forget their silly scheme to change the greatest name that's ever been painted on the front of a fire truck. Back to Station 11 and the Raging Queen …

I patiently listened to my C-shift counterpart explain how they had a lot more pride in themselves than the mouth-breathing B shifters. Their extra measure of pride was at odds with a lot of stuff that seemed centered around B shift. The C-shift captain felt changing the name of our ladder truck might have a calming affect on us. I reminded him that the truck was named the Raging Queen well before either of us had been assigned to the station, and B shift's behavior had nothing to do with the name on the front of our truck. I tried to explain to him that we did the important part of the job just as well as, if not better than, he and his assemblage of C shifters; we just had a different style. He spent the next couple of minutes asking me why there was a bag of chicken bones, dice and dried flowers hooked to the EMS clipboard. I told him we used them on EMS calls because they had a soothing affect on certain patients. His eyes seemed to tear up before he slammed the door and walked out.

After Captain Grumpy Britches stormed off, I changed into my civilian clothes and prepared to leave when a minor argument broke out in the kitchen. One of the C shifters was yelling at one of the B shifters because a bunch of suds were pouring from the front of the dishwasher. The arguing duo stood in the middle of three to four feet of white, soapy bubbles that filled the narrow kitchen. I place this episode square on the shoulders of C shift's obsessive "pride" in managing our station's supplies.

A couple of shifts prior to that morning's blow up, we were cleaning the kitchen at the start of our shift because it looked like A shift had catered an Italian wedding. We discovered we were out of dishwasher detergent, so we used Spic and Span. We finished up the kitchen, turned on the dishwasher and moved on to the next thing. Half an hour later, we noticed suds pushing out of the kitchen. While staring at the 40' blanket of suds, Engine 11's driver wished out loud that the Class-A foam system on his rig made such sudsy bubbles. We quickly determined that Spic and Span was not designed to go into a dishwasher, ordered Engine 11's driver not to put any in his truck's foam tank and made a pre-lunch call to our station's delivery person, requesting several boxes of dishwashing detergent. Our delivery person said he had a full day of work but we would have the soap by tomorrow at the latest. The rest of the shift was pretty uneventful; we had to practice Santeria-style EMS on only a single patient.

The next shift started the same way, but this time the kitchen looked like A shift had served luncheon to the League of Women Voters. We looked for the dishwashing detergent we had ordered the previous shift and found none. After interviewing one of the off-going A shifters, we learned our displeasure with the delivery person was grossly misplaced. Perfect Pete, the A shift poster boy, told us, "When C shift came on duty, they found half a dozen boxes of dishwashing detergent they hadn't ordered. They were seriously pissed off and asked us about it. We told them we had nothing to do with it and said it was probably you guys going around them again. They called the Red Shirt (our delivery person) and told him to come take it away because they hadn't placed the order. Before the kid could haul it off, one of our guys snuck a couple of boxes so we would have some until the supply order comes next week."

I asked if we could borrow a box so we could clean up the mess he and his counterparts left us. He told us, "I'd love to help you guys but it's locked up in my captain's locker and I've got to get going." This started a small station feud. Some anonymous person would leave Spic and Span in strategic places where it would end up in the dishwasher. It could be in the bottom of a dirty cup, directly poured in the bottom of the dishwasher, and in several instances the contents of random boxes of dishwashing agent were replaced with Spic and Span. Every time you turned on the dishwasher, you were playing Russian roulette. This game went on from one season to the next but it had an unintended positive effect. C shift lost focus and forgot about their quest to change the name of our ladder truck.

The years passed, and I thought I was done protecting my beloved alma mater's Raging Queen until another threat came calling—one a little closer to home. I left Fire Station 11 when I was promoted to the rank of battalion chief. I was quite pleased that my brother had enough seniority to take my vacated spot. My joy turned to trepidation when during the course of a routine brotherly conversation he told me he hated Ladder 11's name, and he and the other two captains assigned to the rig had decided it was time to change it. The only problem they were having was choosing a new name. "How about this," I said. "'The truck that transports a bunch of homophobic morons around town.' To be quite honest, with you, my dear brother, I'd just as soon see the truck painted lime green. You can all go to work for the Rural Metro Corporation." I cannot include my brother's response. Suffice it to say it was as spectacular as it was vulgar and obscene. I quickly made calls to anyone I thought could put a halt to this travesty, but in the end the three shifts couldn't come to an agreement on a new name and left well enough alone.

There is a popular saying among new-agers that goes something like, "You never regret the things you do; it's the stuff you don't do that fills you with loss." My decade spent at Fire Station 11 has changed me as a human being. If I had it to do over again, I would change nothing, but would add a few more humanitarian acts of kindness. During that long-gone era, I was endowed with the full powers that come with being the Captain of the Raging Queen. Had I been more aware and in tune with my inner artist, I would have taken full advantage of these inalienable powers and provided the much-needed public service of conducting wedding ceremonies for gay, lesbian and paramedic couples aboard our crimson pirate ship of love. It would have been both glorious and the right thing to do.

Chapter 15
What's Love Got to Do With It?

Do you have to love your job to be good at it? I seem to hear this with great frequency from some of the people I work with. Instructors, teachers and motivational speakers throw this belief around as if they invented it. People seem obsessed with the idea that if you do not love your job, you should immediately relieve yourself of duty and post your résumé on one of the many job-finder Web sites. Please do not confuse the concept of merely liking what you do with loving it; many of these zealots are talking about a love so strong you would drink poison Kool-Aid if your chief, union rep or safety officer told you it was the only way to prove your devotion to the group. Many of these professors of love claim that you cannot soar to the heights of professional excellence unless you love your work. These love-theory devotees fear that the unwashed, non-loving sweat hogs have a negative effect and influence on the righteous and loving members of the clan. It sounds a lot like McCarthyism.

There is a problem with requiring a person to love at a certain level: How do we define an abstract concept, feeling or emotion like love? We have trouble reaching a consensus on the color to paint our trucks, the type of nozzles to use on our attack lines, the style of helmets we will wear, and the six different rank titles that a deputy chief can hold. Up until a few years ago, we couldn't agree on something as basic as how many firefighters should staff an engine company. To further complicate matters, the definition of love (and any other emotion) is highly personal and isn't the best topic for a committee to take on. Can you imagine if the NFPA put together a technical "love" committee? I would buy a plane ticket and a video camera for that meeting.

For me, love is feeling like I'm a little girl in a pretty dress, holding a soft, fuzzy puppy while I eat an ice cream cone on a pier and watch dolphins play in the ocean. At the core of my being, I don't think life gets any better than that, and I must confess my work has never generated that kind of pure, clean, safe emotion in me. Thankfully, there is more than one kind of love. As long as I'm confessing things, I have to say that I do like my job to the point of love, but not little-girl-eating-ice cream-on-the-pier kind of love. I also know firefighters who do not feel gut-wrenching love about their work, yet they don't seem to have any trouble doing an excellent job.

On the flip side of this coin are the members of the fire service who love the job above all else, yet on their best day, their job performance just meets standards. We've all heard horror tales about some of these wackos forcing their family members to call them by their rank. "Excuse me, Captain Daddy, would you pass the peas?" Does an abundance of love outweigh a career of mediocrity? Does superior performance counter a lack of passion? Who gets to decide, the love police?

The issue of loving your job is a whole lot more complicated than driving around with fire-service stickers plastered to your personal vehicle. Many members of the love cult celebrate their love through shift-change rituals and pageants, the use of uniforms and hair regulations. Does love mean you must wear a chrome-plated, phallus-shaped helmet whenever you dress up in one of those ridiculous looking fire-service tuxedos adorned with all that gold braiding? Who in their right mind designed these outfits, Liberace? Love it or not, you couldn't look any sillier if you stuffed a feather duster up your butt.

Another room of the love mansion houses those who feel we should spend every moment of our lives training, pre-fire planning, going on public-education safaris, and performing routine maintenance on our tools, equipment, apparatus and facilities. After a while, this begins to sound like an endless train of busy work to prove your commitment and love. The other end of this scale is occupied by the lazy and incompetent who refuse to do anything that resembles work. This is another lifestyle issue that requires balance. How about this: We train to a level that maintains our skills and standardizes our operations, we conduct periodic pre-plans of the significant hazards in our first-due area, we clean something when it's dirty and we're pleasant to everyone. This philosophy leads to a workforce that provides top-flight service. In the end everything must support going on the calls and doing a good job.

One of the pitfalls of love comes from how we deal with change. We love the things we love because they are the way they are. When those things change we can react in one of two ways: 1) We understand that life is not static and people and organizations change, so we accept it and move on; or 2) We close our eyes, dig in and

129

hope we can stop the currents of change. It has been my experience that the people in the former group flourish; the ones in the latter group will eventually drown.

The happiest and most productive members of our departments seem unaffected by change. This is because most folks want to show up to work—sober and on time—do their jobs, have a good time during the course of their workday and then go home at the end of their shift. These folks don't appear to carry a whole lot of emotional baggage when compared with the angry and frustrated members. A common theme of the disgruntled is, "This place isn't as good as it used to be." These unhappy individuals will profess to have once had an unmatched love for the organization until it let them down. These let-downs include hiring and promoting people different from them, adding non-fire response to our service-delivery menus, holding them accountable for something they did, or passing them on the overtime list because they were on a fishing trip. The nerve.

It is much more telling to watch people's actions than it is to listen to what they say. People who really love 1963 Corvettes love them because they are what they are. A true aficionado of this vintage sports car will cringe anytime they see a '63 Corvette that has been altered in any way beyond how it came out of the factory. A casual car fan may ooo and ahh over this car painted candy-apple red with a brand new 600-horse power motor under (and out of) the hood, but a true devotee will turn away in disgust in an effort to refrain from vomiting. To them, this is no different than moving the New York Yankees to Boston.

This same concept holds true for many other objects de love, including fire departments.

If you want to see how much one of your ranking members really loves their fire department, watch how they deal with the members who they identify as not having the proper level of love and commitment. This is where the issue of "love this job or leave it" really kicks in. We all have our own personal vision and beliefs of what our fire department is: how we perform on calls, the general philosophy and behavior of the members, how well we take care of each other and our stuff, basically our everything. This vision includes the qualities required for something to reach the pinnacle of being perfect. It could be the perfect fire truck, the perfect fire station or the perfect set of gloves. Harder still is developing a set of benchmarks (including love) to measure the perfect firefighter, if there were such a thing. For all practical purposes I could have replaced the word "perfect" with unicorn—neither one is real—but that would have taken me back to the little girl on the pier thing. As hard as A-Shift tries, perfect doesn't exist.

I have never seen a performance evaluation that contained a category for love. One of the reasons we don't evaluate love is it is very difficult for a supervisor to measure an employee's emotions. The most common enterprise that begins with taking an oath to love is marriage, and half of all marriages end in divorce. A few months ago, I sat through a management class. It seemed that every 10 minutes the instructor would come back to the concept of loving—loving your fire department, loving your station, loving your fellow workmates, loving training and maintenance, loving all the calls you go on, loving, loving, loving. After a while he started to sound like an interior designer talking about this season's color pallet and fabrics. All the love-speak was catchy and new age, but none of it provided any insight regarding how an officer could be a more effective supervisor. Holding true to form, I daydreamed myself back to an earlier time, when despite my best efforts, I was assigned to a committee charged with finding ways to improve our department's training. One of the principles of this group made a very profound statement that has stayed with me ever since.

The committee of response chiefs, training academy staff and field captains was convening for our 1,000th meeting. This one was to discuss the organizational expectations for probationary firefighters. This discussion included the qualities required to become a new recruit. The banter was all over the place. "We should only hire the young. We should only hire people having 10 years of work experience. We should only hire cowboys. We should only hire people with college degrees. We should only hire people who speak six languages. We should only hire people who can benchpress 400 lbs. We should only hire the psychic..." This went on for a painfully long time until one of the captains sitting in the back of the room spoke for the first time. "The only parameter I use to determine whether or not a new employee will become a good firefighter is if I can stand spending a 24- hour shift with them." This comment angered many of the social scientists in the room because it can be difficult to measure what makes a person tolerable for long periods of time. I feel this is true because we make these judgments based on a mix of neuro-signals we receive from our higher brain, our emotional centers and our reptile brain. In many ways, it's like trying to describe the taste of chocolate.

How do you define a subjective area like social skills and develop an evaluation process that rates a member's commitment, trust and love? That is one of the major reasons you don't find these performance measurements on employee evaluation sheets. Try defending this type of rating system in a grievance hearing or open court. "We suspended him, your honor, because he didn't love us enough!" Case dismissed.

As if all this wasn't complicated enough, it is commonplace to have a wide range of opinions within the organization about the same person. One day a firefighter raves about what a great officer Captain White is. The same day another person goes off on a rant about that same Captain White being a double-dealing, incompetent douche bag. The heavy hitter in this discussion is your boss, because they're your supervisor and they fill out your annual evaluation. This can be a slippery slope: What happens if you're the employee of the year but your boss happens to be a douche bag and rates you poorly because you had the nerve to disagree with their position about Adolph Hitler's excellent management theories?

I loved working at certain stations with certain people. There were other people I worked with and other parts of town I worked in that I didn't love. We all know people who were excellent firefighters but upon promotion became lousy officers. No amount of good firefighter skills could make up for their poor officer skills. In most cases the love they had for their job stayed the same. I suppose love alone doesn't make a good officer.

Another problem with establishing a required love level is people tend to expand their definitions outside the boundaries of your standard work requirements. Keep your shoes polished, throw out the trash, clean the head, and you're on the road to fulfilling your employer's love requirement. But if you also practice witchcraft and sorcery on you days off, does that equal a poor moral code and cause you to earn job-related love demerits?

It's important to enjoy what one does for a living. It is also important to be competent at what you do. It is a lot easier to spend a career doing something you enjoy (or love) than it is to dread work everyday for the next 30 years. It is another matter altogether for the employer to demand that you love the organization. We are not a church. People are capable of feeling a full range of emotions. One day you love it, while the next day you want to break it into a thousand pieces. These are the ups and downs of life. Much of a person's perspective and emotional state is based on where they are, who they're with and what they're doing. The Pandora's Box of job love is bursting with many complicated issues. Some days I feel in my heart of hearts that life would be a lot simpler if I were a little girl eating ice cream while looking at the ocean.

Chapter 16
Oh Christmas Tree

My first fire was exciting and tense. It was also a matter of life and death—my life and my brother's hung in the balance. I was a senior in high school, and my brother was a freshman. The whole family was starting the day out together, and Dad was getting ready to go off to his fire chief job. Before he left, he gave my brother and me explicit instructions to get rid of the 12' Christmas tree that occupied the living room. My brother and I had put this chore off for quite some time; Valentine's Day was only a week away. Dear old Dad was rapidly losing his sense of humor about our lax approach to Christmas clean-up, so my brother and I figured we would remove the bomb that occupied the corner of our living room when we got home from school. Mom and Dad both left for work, and my sister boarded the bus that would deliver her to the elementary school down the street. Then my brother and I decided that our chore was too important to put off. For safety's sake, we stayed home from school to remove the menace from our home.

The first order of business was to eat. One needs energy when taking on arduous and life-threatening tasks. The two of us drove to the local Denny's and discussed our tree-removal strategy over a couple of Grand Slam breakfast specials. Our father had instilled the importance of proper planning and management to us at an early age. By the time we finished breakfast, we were ready to take on the tree.

When we got home, we started on our chore. The first step was to remove the ornaments and lights. We carefully and lovingly removed Mr. Snowman, Santa Claus and the Three Wise Men (which had been skillfully handcrafted out of yarn in the Philippines), wrapping them in protective tissue and putting them back into their boxes. Next off were the lights and garland. We had strung popcorn on kite string. (My mother had used the leftover popcorn to make red and green popcorn balls. I still had some of this Yuletide confection wedged in my molars.)

After we had the tree stripped down to its natural, albeit dead, state, we stuffed it into the fireplace. This was the core of the plan we developed over breakfast. We did not want to hassle with taking the tree to the dump and were too civic minded to just dump it on the side of the road. The crux of our plan was the soundness of the fireplace and flue.

Six years earlier, I had helped my father and another psychopath build our family home. My father and his building buddy were advocates of over-engineering. The fireplace flue was a vault. Large 1½" thick clay flue tiles lined a chimney made of grouted 8" x 4" x 16" cinder block. The space between the flue tiles and the chimney was then filled with enough concrete to make an average family-size swimming pool. My brother and I reasoned that since the structure could serve as a launch tube for a Saturn rocket, a Christmas tree should be a walk in the park. We stuffed the tree up the fireplace, big end first. We had to do it this way because the bottom of the tree was at least 6' in diameter. This took about an hour, and it was not as easy as you'd think. We were both impaled with splinters, bleeding and pissed off by the time we got the tree into its final resting place. I recall my brother and I took a 5-minute break in the middle of this disaster to have a fistfight. Things were not going well, and they were about to get much worse.

I'm sure many of you are wondering why I'm not writing a story about my first fire as a fireman. The simple reason is because in all actuality, this was my first fire, and it was much more exciting than the first one I had as a member of the Phoenix Fire Department. I also had a lot more riding on this one.

After my brother and I packed the tree into the firebox, we lit a single match, tossed it in and closed the heavy brass screen. Within seconds, the heat drove us across the living room. It was the first time I had ever witnessed radiant heat and convection currents. Our clothes were hot. My brother had a look of abject fear on his face. I'm sure I wore the same look. We were convinced that we were in the process of burning down the fire chief's house. I looked at my brother and told him the whole thing was his fault.

In the span of just a few seconds, we had gone from a single lit match to nuclear fusion. As if the heat and blistering flames weren't enough, they were quickly joined by small pinesap detonations. It sounded like mortar shells going off. The heavy fireplace screen kept the incendiary embers from flying out and setting the living room ablaze. The fire was burning so bright that it hurt our eyes to look at it.

After a couple of minutes, we fled the interior and went out front to see how the chimney was doing. My brother and I stood hypnotized and helpless in our front yard. We had only lived in our home for six years. There couldn't have been that much soot, creosote and unburned fuel stored in such a new chimney flue. But there was 40 feet of fire blowtorching out the top of the chimney. My mother and father's anniversary was the following week, and my brother and I were in serious peril of burning down their dream home. We couldn't blame this on my sister. Our parents were too smart.

The entire event did not last 5 minutes. As my brother and I stood there weak-kneed, the fire quickly died. We ventured back into the house to find no more damage than a light haze and what felt like a 25-degree temperature increase in the living room. The fireplace was devoid of any signs that anything had been burned in it, let alone 1 million BTUs of last year's holiday tree.

My first fire left me both invigorated and emotionally drained. That day I learned how important it is to start your day off right with a nutritious breakfast, to over-engineer safety devices and to dispose of flammables properly.

Chapter 17
Bridge to Nowhere

Safety may as well be a bat we use to beat one another with. In some cases, I think we've taken the safety concept too far, or at least we may have distorted it on some level. But it's pretty tough to argue against any safety-related topic without looking like an idiot. In the most practical sense, safety can be a challenge because it encompasses such a broad set of practices, behaviors and attitudes. These safety concepts include very specific and easily defined (and enforceable) principles, such as "always wear your seatbelt," or very vague ones, such as "don't commit unsafe acts." (Once upon a time a safety officer lectured me about this.)

Every human act has a safety component, and one of the beautiful things about safety is that it lends itself to personalizing. Each of us can develop and maintain our own safety idiosyncrasies and pet peeves. People who are obsessed with safety tend to lead very neurotic and constricted lives. On the other hand, the people who have a total disregard for safety keep the prosthetic industry thriving and die premature deaths. Life is all about balance. Any fire-service safety discussion takes on a whole new dimension because of what we do for a living. I have always felt it was odd, if not schizophrenic, for us to discuss "safe" operations conducted in and around burning buildings. The following episode is a perfect example …

My department was having a quarterly ops chief

training session about fires in buildings with bowstrung roofs. The class began with a forensic structural engineer presenting several case studies where some local buildings with bow-strung construction had simply fallen down due to age. Each building had woken up completely normal, but sometime near bedtime they broke and fell down. None of the buildings had caught on fire, hosted gas explosions or had meteors fall on them. In every case, the tenants showed up to find a large pile of mysterious rubble that used to be a perfectly good building. According to our scientist instructor, they had all experienced a "catastrophic longitudinal crack of their second truss member" and fell to ground.

So the front end of our class established that it isn't safe to shop in some of these buildings, even under ideal conditions. They can collapse without warning, fire or no fire. In the next part of the class, the instructors had us review a set of fire-simulation pictures showing buildings with bow-strung roofs. The instructors then asked us to identify the correct fire attack strategy based on fire extent and location. Keep in mind that these were not yahoo instructors. Both instructors were very good at what they did, but this didn't change the fact that the two parts of our class didn't match up.

To be perfectly honest, I think all of us have a skewed perception of safety instilled in us during our formative years. Most of us came to the fire service cooked a little differently, and this personality trait is taken to a higher plateau when the incumbents take possession of us. My first field assignment was on a truck company.

A regular part of our day was spent training, and this is where I learned many of the task-level details of truck work. During my first month, we had several fires where we vented roofs, pulled ceiling, forced entry and did all the other stuff a ladder company does at the scene of a structure fire. One day we caught a working house fire while we were out, and we ended up being the first-arriving unit. I got off the truck and grabbed a chain saw and an ax. My captain met me at the front of the truck and told me to put the tools on the ground. This was new territory for me; we had always arrived after the first engine or two were already on scene, and we performed traditional ladder work every time. I dropped the tools and asked my fearless leader what he wanted me to do.

He replied, "Mask up. You and me are going to wait right here for our engine to show up, and we're going to swoop their line." I thought this was one of the greatest ideas I had ever heard.

I was giddy listening to the siren of our engine a few blocks away as I stood on the curb with a free-burning house at my back. My joy quickly turned to dismay as my captain and I watched the first-due engine company slide (as in a speed-induced fishtail) around the corner. The senior firefighter on the engine was standing on the soft suction tray on the side of the truck. He was fully turned out and masked up, with his arms through the loops of the 1½" attack line. He had the working end of the line draped over his other shoulder with the nozzle in his free hand. As soon as the truck slid to a stop, he pulled the hose bed, and the engineer charged his line as he disappeared into the smoke. My captain shrugged his shoulders, and we headed to the roof to do ladder work.

After the fire, I asked the tray-riding firefighter why he did what he did. He told me he heard our captain give an on-scene report and knew that we would probably try to take his attack line, so he threw his mask on and put himself in a position where we couldn't "steal his shit." This taught me a significant lesson early in my career. They didn't teach me everything I needed to know in the training academy. I learned lots of unsanctioned task-level cheats, techniques and philosophies during my formative years:

- Three Nomex hoods sewn together will protect you better than one.
- If you arrive on the scene of a working fire, and you and your partner are both rapidly donning your SCBAs in an effort to be the first one to the nozzle, throw his gloves under the rig; you will win.
- If you let the smaller firefighter have the nozzle, you can get right behind them, lift them off the ground and use them as a heat shield. Sometimes this is better than having the nozzle.
- It really scares the initial attack crew when the ladder crew climbs down a hole they cut in the roof and charges toward them.
- It's a lot faster to remove roof tile with a deck gun than by hand.
- Lying on your back with the nozzle fully opened after the room flashes over isn't as effective as being out of the room.

- When your protective gear begins to smolder, you should quit advancing. It is much better to have the roof collapse while you are forcing entry than having it collapse 1 minute after you get inside.
- The biggest difference in all of life is the difference between dying and almost dying.
- On the other hand, people who have lots of near-death encounters tend to die early.

For better or worse, I believe our proclivity for action and our highly competitive nature represent the true spirit of the American Fire Service. It's part of our 300-year history, and a lot of the actions we take today are based on a model we haven't questioned very much during the last three centuries. Sometimes when I reflect on my career, it occurs to me that my elders were trying to kill me. I was a complicit participant and had the time of my life, but many of our fireground operations could only be described as insane and pointless. Many of our actions occurred completely outside the bounds of our risk-management plan. It's almost as if we were combatants in some sort of pagan fire right of passage.

Several years into my career, I worked at our downtown firehouse. A ladder company and a couple of engines ran out of this station. I was assigned to one of the engines. The ladder captain fancied himself as a hybrid beatnik-intellectual sort of fellow. The type of guy who wore Birkenstock shoes before they became the official footwear of the Lesbian Nation. He was of a slight build, had male pattern baldness and wore John Lennon glasses. Each member of his crew outweighed him by a minimum of 100 pounds. This particular officer was a training fanatic. Every shift he would drill with his crew for hours, often times spending eight hours a day doing hands-on training. This ladder captain was expanding the training curriculum beyond what any clear-thinking individual would consider standard, safe or sane. There is only one true finale to this type of situation: mutiny.

The training session that finally caused the crew's revolt is as ludicrous today as it was all those years ago. The five-man crew was conducting their daily drill on the roof of a six-story building. During the course of the drill, the captain had the crew loft a 35' extension ladder up to the roof. The two firefighters went down to their rig to retrieve the ladder, while the engineers fed a rope over the side of the building. The ground guys tied the ladder off, and the roof guys pulled it up. After the firefighters made their way back up to the roof and the crew was reunited, one of them asked their officer why they would want a 35' ground ladder on the flat roof of a mid-rise building. The ladder officer explained that he'd been contemplating safe escape paths from the rooftops of high-rise buildings. After pointing out a 10-story building with an exterior fire escape directly across a narrow alleyway, the captain explained the day's next training evolution.

The crew fully extended the ladder and lowered it across the 25' to 30' gap, resting it on the exterior fire escape across the alley. The ladder became a bridge over the 60' high chasm. Placing the ladder in this fashion required the three largest members of the crew to butt the ladder while it was lowered, and nothing about the operation could be termed as safe or sane. The fully extended ladder was long enough that it could be butted against the wall of the 10-story building so it wouldn't move from side to side (and possibly slide off of the fire escape railing and make the six-story fall to the street) while the next piece of this madness continued. The captain's explanation for this improvised ladder bridge between the two buildings was based on a make-believe scenario involving a fire on one of the lower floors of the building they were standing on top of. The occupants of the building were unable to exit the structure via the elevators or stairs and had to seek refuge on the roof.

The captain posed this question to the bewildered crew, "How would we rescue these people from the roof?" The answers ranged from using aerial devices, leaving them on the roof while we put the fire out, helicopters, catapults and harpoon guns—in that order. The captain kept playing the "what if?" game and disqualified all of their options with a variety of conditions that would have made these options impossible (no access, mechanical malfunctions, etc.). The only solution was a ladder bridge. The next step was to have the crew test the theory.

The first man selected to go over was one of the engineers. He was 5'8", 250 pounds and had the stunning good looks of Sebastian Cabot, the actor who played the popular Mr. French butler character on the "Buffy and Jody" show that we all grew up with during the 1960s. The truck rope was used to make a harness and lifeline for the engineer. While preparing this safety precaution,

145

the crew began to voice their doubts about the captain's ladder-bridge concept. The crews concern grew as they tied up the engineer like a trussed turkey right before the captain ordered him across. One of the firefighters was placed in charge of holding the other end of the rope and providing slack as his fellow crewmember crawled across.

As one might imagine, ground ladders are designed with a climbing angle in mind; they weren't intended to be used as scaffolding (or bridges). Butted on the other side by two firefighters and jammed right against the building across the street, the ladder didn't have any side-to-side movement, but there was motion.

Every time the engineer advanced and shifted his weight, the ladder bounced up and down. After he advanced five or six feet, he began to shout obscenities at his captain. This was all the rest of the crew needed to become unhinged.

The firefighter holding the other end of the rope had figured out that if the worst should happen, he would be pulled over the side of the building with his partner, all the way down to the alley below. He chose life, threw the rope to the ground and called for the engineer to come back. This threw the captain into a tizzy. He ordered the firefighter to return to his post and man the safety rope. The firefighter told his captain to pack sand in his ass and to man the rope himself. While this was going on, the two other crew members assisted the engineer back to the safety of the roof. For the next 30 minutes, the crew shared their feelings with one another. When they got back to the station the captain went home sick.

Each of us has our own personal safety code. These beliefs range from flat-out insane all the way to paralysis. Captain Birkenstock's safety ideas resided in the Evel Knievel wing of the nuthouse. During this same era, our department's safety officer held a completely different view of what was safe and would have had Captain Birkenstock committed and studied had he found out about the improvised bridge drill.

It is easy to sit around an air-conditioned office and pontificate over what is and isn't safe. It is quite another thing to have to manage scene safety when the building is on fire. Safe operations are the result of well thought out SOPs, effective training programs, and operating within an incident management system that connects all the incident participants. These hazard zone pieces are bolstered with accountability systems, safety officers and RIC organizations in an effort to make the workers safer. The real problem is we have all been to incidents where we had multiple layers of safety in place and firefighters still got injured (or even worse, killed). On the other hand we've all been part of operations that were so safe, sound and effective they are best described as elegant. In many cases these incidents had no dedicated safety officers, all the passports stayed on the dash, and the RIC component was covered by maintaining a tactical reserve. Everyone just did their job.

My editor has told me on numerous occasions that my articles should make a point. But for the life of me, I can't figure out the jumble of firefighter safety. We have dedicated safety officers, PASS devices on our SCBAs, accountability systems, rapid intervention, safety rules and regulations, etc., but on a fundamental level it all seems to somehow miss the main point. Real deal safety is part of every task-, tactical- and strategic-level action we take. It doesn't work very well when we try to distill "safety" out of standard operations and assign it to the safety division. Our work can be so fast and unforgiving that if we miss something the first time, we don't get a second chance.

Chapter 18
Goat Boys

Many Christmases ago, my brother received a bow and quiver of arrows from Santa. This is not the smartest gift for a 12- year-old boy who has a running jihad with the next-door neighbors. The blood feud between my brother and the next-door people began several years earlier when on a lazy summer day, he and his best friend took their beloved pets for a walk to the park. My brother's pet was our family dog, a 220-lb. mastiff. His friend's pet was an angry Billy goat with horns painted shocking neon pink. When the city irrigated the park behind our houses (flooding it with about 6 inches of water), the boys would play gladiators and chariots by rigging harnesses and small sheets of plywood to their pets. It was quite a sight to watch outlaw 12-year-olds surf behind a large dog and a bucking goat.

Our next-door neighbor was a surly old woman who didn't cotton to young 'uns frolicking in irrigation that her tax dollars bought and paid for. She attempted to shoo the boys out of the park, which led to an unfortunate encounter between her ass and a set of pink goat horns. The old house frau called the police, earning the scamps a stern lecture regarding the hazards associated with playing in irrigation water of an undetermined quality and the future consequences of assault and battery with a goat.

All of us have a nasty old lady that lives on our street, and most people don't pay much attention to her constant grievances—we wish she'd just get a life. The set of parents in this fable were no exception; they told their sons to stay away from the nasty lady next door. They did urge the boys to keep better control of their animals, because the goat just as easily could have ass-rammed a kind and decent person.

Over the next few months, the biggest issue between the lads and the crazy lady was the ever-decreasing length of her horses' tails. She was convinced my brother and his friend were sneaking over at night and grooming her ponies. The mystery came to an abrupt end when she came home with a pair of Doberman pinscher roommates for the ponies. All was calm until that fateful Christmas, when my brother received his treasured bow-and-arrow set. This was not a kid's toy with rubber suction cup arrows. It was a long bow with steel-tipped arrows. (The image I still associate with that Christmas is my brother holding the bow behind him, placing it between his legs, hooking the end of the bow around one foot, utilizing his opposite leg as a fulcrum and using his full might to bend the other end over in order to string the weapon.) For the rest of our Christmas vacation, young Robin Hood's quiver grew increasingly sparse, as evidenced by all the arrows sticking out of palm trees 50 feet in the air, along with the misfires that landed God knows where.

The few remaining arrows that led to the bow's permanent removal from my brother's arsenal had been fired through the next-door neighbor's bathroom window, impaling themselves four to five feet above her commode. The irate, mumu-wearing, curlers-in-her hair neighbor showed up at my parents' front door with a fist full of arrows, foaming at the mouth and ranting in a language that none of us understood. My parents calmed her down and assured her that the issue would be handled without passion or prejudice and would not be repeated. Under serious examination, my brother quickly broke and admitted his misdeed. The confession so surprised my parents that they were at a loss over stern discipline and settled on the destruction of the bow. My brother feared that my parents were only toying with him and assumed there would be a future flaying. To date, it is the only time in the history of our family when one of the children looked forward to going back to school.

In the end, the thing that saved my brother was taking personal responsibility for something he did. Thirty years later it's pretty easy to sit back and think he really didn't have any other choice—after all, he was the only 12 year old in the neighborhood who was running around shooting green, steel-tipped arrows with yellow feathers. The other mitigating circumstance was the neighborhood's contempt for the victim. This was a woman who went on a fruitless crusade to have a little boy's beloved pet goat destroyed, and many of the residents felt she had it coming. I think it was the first time my brother ever admitted to wrongdoing, and it threw my parents for a loop. All previous incidents ended with name, rank and serial number or a flat-out lie that it was my sister's fault. (This became my brother's standard response, which would elicit a standard response from my parents.) The act of contrition, combined with the mitigating circumstances (i.e., a goat-hating hag) turned the tide. Life would be much simpler if adults acted like my 12-year-old brother did back then.

One of the problems plaguing the full spectrum of society is no one takes responsibility for their actions. In fact, this may be the mother of all problems and the root cause of all other problems. The problem ranges all the way from the President of the United States ("I did not have sex with that woman," or for you liberals, "We will find the WMDs"), all the way to petty criminals who blame society for their lack of opportunity and the general insolvency of their wretched lives.

Several months ago, we had an incident where a firefighter operating in an aerial platform above a defensive fire got burned. This injury occurred for several reasons, mainly because the platform was operating in the wrong position and the firefighter wasn't wearing the proper protective gear. The entire episode was caught on videotape (from multiple camera angles), and since no one was disabled or killed, and this wasn't the only out-of-balance act that occurred at this particular incident, we shared the incident critique and the lessons learned with our entire department (and anyone else who would listen; we did a workshop on this incident at our department's annual incident management symposium).

A week or so after the flaming-bucket incident, several chief officers were discussing the event at the fire chief's monthly meeting (the same man who broke my brother's bow-and- arrow set into dozens of little pieces with his bare hands). I sat in stunned amazement as I listened to response chiefs making arguments that the fix for the safety violations at this incident was somehow training related. After the first guy offered up the training antidote, some of his counterparts began to chime in with similar comments. Having felt like I just ingested a platter full of brightly colored hallucinogenic rain forest frogs, I asked a question befitting a deranged homeless person: "Does this mean we're going to go out and train all of our members on how to put on their gloves?" The fire chief, who was enjoying the discussion, shot me a polite reply that really meant shut up and urged the young, intelligent and vibrant battalion chiefs to continue. They were ready and willing. "The sector officer on the backside of the building had a crew advance an attack line into a position they shouldn't have been in, and a falling wall knocked one of the firefighters senseless. No one ever trained us what to do when crews take offensive positions at defensive fires." I wanted to crawl into a dry cleaning bag and take a nap.

Most of the unsafe acts I have been party to weren't the result of insufficient training. In fact, it was quite the opposite. I was able to survive unsafe actions because of the quality of my training, along with an abundance of sheer luck. Our department has taken a very healthy and positive organizational approach to sharing the lessons learned when one of our members performs unsafe or non-standard actions by incorporating those lessons learned into department-wide training. In many cases, these events are the result of acts most of us have done on an almost routine basis and served as a vivid reminder of why we call it a hazard zone. Analyzing and sharing these "near hits" has become the most effective way I've seen to improve firefighter safety.

Despite the organizational mileage we get out of these events, we do not execute our members when they make mistakes. If that was our practice, none of us would still be here. Admitting that you make a mistake is the first step in keeping it from happening again. You have got to wonder when a fire chief says, "We wouldn't have changed a thing…" after a firefighter dies in the line of duty. Do you think that chief's firefighters look at one another and wonder who'll be the next one to bite the dust when they do it exactly that way again?

I have been involved in my share of fire attacks that didn't have any resemblance to our SOPs. I recall standing with the assembled group after some of these tactical misadventures, listening to battalion chiefs gloss over serious errors, commend poor attack-line positions and late salvage work, only to become irate when one of the firefighters had the nerve to wear a uniform shirt with cracked stenciling on it. All of the workers recognized their mistakes and always thought less of the boss when he chose to ignore them. The best bosses I've worked for were honest when critiquing an operation without making us feel stupid when we did something wrong.

One of the benefits of the training programs our department has implemented over the last four or five years has been the standardization of structural firefighting. This has created a system where the post-incident critiques are based on the same systems we use in training and actual incident operations.

This has shifted our after-incident review from a rank- or ego-based critique to an approach grounded in, "What did you have when you arrived on scene?" and "What did you do based on what we agreed on over the last four or five years?" This has brought us closer together in an operational sense. Senior members of the command staff have remarked that at our department's recent critiques, all the incident participants appear to have been at the same fire, whereas in the past the group of incident accomplices sounded like they were all at different fires.

A few weeks after the fire chief's meeting, I sat in on our department's critique of the flaming-bucket incident. Many of the participants offered a variety of explanations regarding why and how things got out of whack. It sounded a lot like doomsday psychics trying to explain why their end-of-the-world prophecies didn't come true. Deflection abounded until the sector officer operating where a crew advanced an attack line up to the doorway of a fully involved defensive fire spoke up. She told her story and ended it with, "When I came around the corner and I saw where the crew was, I knew they shouldn't have been that close to the building, but I didn't have the balls to tell them to move."

I learned several things from this critique: 1) It was the first time she had ever been a sector officer; 2) We should train officers on how to handle situations where crews are operating in unsafe positions; and 3) She had the biggest balls in the room. There are a couple of things at the heart of this matter. We are all human and like to be recognized and commended when we do something right, but it is more refreshing when someone who makes a mistake takes the credit for it.

Chapter 19
Spare the Rod, Spoil the Fun?

When we all scream for discipline, we really want punishment. On one level, we believe we can punish our way to perfection. This is not a new concept; it's been with us since long before a famous group of fanatical discipline advocates wrote the Old Testament. On a deeper level, I think punishment is somewhat voyeuristic. We feel fortunate it's not happening to us, like the little pig watching from the safe confines of his brick house as the big, bad wolf consumes his siblings. Human beings have a tendency to turn the severe punishment of others into spectacle and entertainment, as evidenced by public hangings, stonings and floggings.

During Charles de Gaulle's reign as the president of France in the 1960s, he commented, "How can anyone govern a nation that has 246 different kinds of cheese?" Imagine how difficult it would be to develop and enforce a comprehensive set of rules that covers every situation and problem. This becomes even more daunting when you consider that many four-person engine companies have major difficulties agreeing on the two meals they'll eat during a 24-hour shift. Multiply the simple dilemma of what's for lunch by the entire population of your department, and imagine the complexity of defining acceptable standards for behavior, grooming, conduct, performance, etc.

When a group's stern punishment advocates deem that one of the members deserves strong punishment for a rule infraction, it can have far-reaching affects on group dynamics. In most cases, these pleas for severe punishment are part of a never-ending string of hysterical rants from the cheap seats. The mojo can run the other way when the outcry for whipping-post-based personnel solutions sets off an organizational stampede akin to screaming "fire!" in a crowed theatre. The desire to see one member of the group caned within an inch of his life can become more infectious than the flu. This is a natural element of the human condition, and when amped to its highest level, it becomes the foundation of all lynch mobs. Burning witches at the stake is always more fun and fulfilling when we do it as a community.

The challenge of discipline and punishment is ensuring it produces the desired behavior and maintains general order. It doesn't make any sense to flay the workforce simply for the amusement of the bosses. Many organizations claim to be well disciplined because they are always reprimanding the workers. On the surface it may appear they are running a tight ship, but does frequent and firm discipline equal high efficiency? How do you evaluate the effectiveness of all this discipline? Even more important, do the rules and regulations fall in line with why the organization exists in the first place? Daddy may claim he only hits mommy because he loves her so much, but does that make it true?

Let's take a look at past strategies for castigating the workforce. One of my early customer-service role models was one of my department's first paramedics. For the sake of decorum, I will refer to him as "Stalin."

When I met this early pioneer, he was a veteran firefighter of eight years, having been a medic for five of them. I was a fairly new firefighter when he explained the basic concept of emergency medicine to me: "Kid, we may be first when seconds count, but that will never change the fact that after midnight, the patient becomes the enemy." (An interesting side note to this story is back in those days, station captains filled out and maintained the station logbook. Each shift began with the officer recording who was on duty and which rig they were assigned to. Stalin's captain would add a symbolic third name to the roll call of Stalin's rescue company so it always read like this: "Stalin, Deaver and Death. Deaver was Stalin's flesh-and-bone medic partner.)

Stalin was well known as a competent medic, but his true talents lie in being a spectacular rule breaker. He had a total disregard for all authority and was always at odds with some medical director, doctor, captain or chief. He also nurtured love/hate relationships with nurses. He was very creative; inventing many phrases and plays on the English language. He was the first person I ever heard use the same profanity as a verb, adjective and a noun in the same sentence: "Fuck the fucking fuckers." Memories of my early relationship with Stalin caused me great confusion when I worked for him 10 years later. I was a new officer and had just taken a spot at a station in Chief Stalin's district.

My first shift began with Stalin and my battalion chief calling me to their office for a sit-down. The get-together began with the standard "welcome to the battalion" speech, and then quickly turned into a scolding built around vague rules and possible punishments for not complying with the Stalin's dictums. I felt like I was trapped in a "Twilight Zone" episode. I really thought he was pulling my leg and made the mistake of saying, "Come on, you of all people have got to be kidding." This brought on a 20-minute lecture about the length of my hair and my indifference toward organizational rituals. (It's weird when one considers that the most enthusiastic advocates of the temperance movement are often reformed sinners and ex-drunks.) My dressing-down ended with Stalin making a sweeping gesture at a shelf full of volumes, SOPs, rules and regulations behind his desk. "Someone took the time to write all of these rules," he said. "I think it is your duty to follow them and my duty to enforce them."

Over the course of our meeting, he ate an entire box of Ritz crackers, smearing each one with mounds of peanut butter and jelly. During the 60-minute meeting, he consumed enough calories to turn a bathtub full of water into steam. For the next five years, Stalin was the head honcho of my battalion. He focused on uniforms, paperwork and the nebulous concept of maintaining good order. Stalin never seemed too concerned with actual job performance; in fact, the employees he held up as examples of top-flight firefighters and officers had the shiniest shoes, shortest hair, kissed the most ass and had the most citizen complaints. Many of them had also taken flooding their hose beds to a professional level. During that time, the saddest I ever saw him was after the Berlin Wall fell. "Those Communists had a knack for running a country," he said.

I am not advocating complete anarchy. We get paid to do a job and need to maintain a workforce that can deliver service when the lights come on. Our rules should be simple and straightforward, and you shouldn't need teams of personnel Nazis, union reps, lawyers, arbitrators and judges to figure them out. They should also be designed around the service we provide. Because we routinely deliver service in and around hazard zones, our safety rules are just that—rules (as opposed to suggestions).

These safety rules must be crafted around the stuff that can injure and kill us, things like riding and driving, using the proper protective gear and operating as part of a company. Real-deal discipline revolves around never wandering off a attack line when operating inside a burning building.

On the non-hazard-zone front, our rules and regulations must reinforce getting along with one another and taking care of the customer. When we join a fire department, we become a member of the organization. The organization must create an atmosphere that causes us, the employees, to treat one another as well as it expects them to treat the customer. It doesn't make very much sense for the managers and bosses to club the employees at regular intervals and package it up as some kind of leadership program designed to instill efficiency. Most people who get clubbed are looking for some kind of payback as opposed to going out and being nice to their diverse and unwashed customer base.

Discipline ought to be corrective, not punitive. When someone makes a mistake, the process should be designed to correct the problem and improve performance. A lot of discipline happens because the person who owns the power to wield it decides they aren't doing their job unless they are dishing out harsh punishment. I learned one of the most profound lessons of my life when I got caught up in this dysfunctional side of discipline. It all started one day at my mom and dad's house. I was sitting in their living room, waiting on my brother, when I innocently picked up a well-worn paperback book, "Dare to Discipline," sitting on the table next to me.

I leafed through the pages and quickly determined that it was the work of someone who enjoyed spanking children. While I read about how sparing the rod spoils the children, my sister and one of her associates came into the room. The associate was a longtime family friend. She walked over and asked for her book, which I was only too happy to be rid of. The two of them explained to me the deep mammalian need our children (or was it cubs?) have for a clearly defined parent-child relationship. The cornerstone of this parent-child bond is frequent discipline meted out by the parents. Apparently this leads to well-defined social order within the tribe and an abundance of family harmony. I replied that my sister and her friend had all the mothering skill and nurturing ability of swimming-pool chemicals and left. Only later would I find out that these two harpies had planted a diseased seed in my brain that I would soon have to deal with.

Two months later, I was at home with my 2-year-old daughter. We were cleaning her room when we got into some kind of disagreement. We argued for a couple of minutes, and I attempted to end it with the standard parental statement, "Because I am the parent." She wasn't having any of it when I flashbacked to my "Dare to Discipline" class. I then set about establishing a strong parent-child relationship and gave my Little Princess an affectionate swat on her behind. This was the first time I had ever ventured into defining the familial relationship with loving violence, so I really couldn't have known what was to follow. My Little Precious wore a look of total surprise for a full count of three before she told me never to do that again, punctuated her statement with a punch to my groin and asked, "How do you like it?" This caused me a certain amount of discomfort, colored with a small measure of rage, leading to a slightly firmer whack and a reaffirmation that little girls were not ever to punch daddy in his balls.

If my first loving, bonding whack caused my Little Angel to turn into a self-determined woman, my second spanky-spanky turned her into Satan. Her eyes rolled back in her pretty little head, and the scream that blew past her raspberry lips broke the windows in her Barbie playhouse. Her arms looked like airplane propellers as she struck me with great velocity and frequency. I took the only action available to me: I ran away and locked myself in my bedroom. As I stood behind the locked door she was attempting to beat her way through, I cursed myself for listening to my sister and her Jezebel companion.

My baby's incoherent wolf howls slowly turned into

understandable English as she screamed, "Open this door and let me in!" Taking a page from the Catholics (who are much better at manipulating people's behavior than my sister and her she-beast friend), I shrieked back, "You can't come in until you settle down, apologize for hurting daddy and give me $100." When my daughter's rage slowly settled to low-level crying, I opened the door and let her in. We both begged one another's forgiveness for the foolish incident and celebrated with martinis and milkshakes. Some good came out of the episode because for the next few years whenever my daughter started to get out of sorts I would threaten her with, "Behave yourself or daddy will lock himself in his room."

In the end I feel that punishment and violence generally just beget more punishment and violence. There are some organizational exceptions to this rule, but I have gone on long enough and will save the last word for a young firefighter who felt there wasn't enough punishment being administered to the members of our laid-back fire department. One day he was lamenting before a group of us that one of his colleagues had gotten away with a minor rule infraction with only a brief lecture. I don't recall the specific transgression, only that it wasn't that big a deal and telling the person not to do it again seemed to be the appropriate course of action. After listening to this young firefighter for what felt like a lifetime, I finally interrupted and asked him, "Why do you even care about this petty bullshit?" His reply cut to the heart of the matter, "I did almost the same exact thing a year ago and ended up being reprimanded, and a letter of counseling went into my file." I asked him what kind of moron would go out of their way to bust a guy's balls over such a trivial thing. I wasn't surprised when he told me, "Chief Stalin."

Chapter 20
George W. Bush's Command System

My first attempt at this column began with a list of all my department's endeavors to obtain the Holy Grail of incident management—the coveted "NIMS compliance." This included defining the five types of incident operations described in NIMS and the National Response Plan, and a review of the professional qualifications for the incident-management positions required by NIMS. Rather than bore you with a bunch of bureaucratic B.S., I will take this opportunity to howl at the moon and blow off some federally generated steam.

Every single fire department in the US of A must be NIMS compliant at some date that keeps getting moved back. Once the final date is final, the feds can withhold all federal funding from states whose counties, cities, towns and villages refuse to comply with the NIMS mandate. This not only includes federal grants and other federal giveaway programs, but also highly coveted federal highway funding, which is serious money. Our federal NIMS masters are a freaky 12-headed hybrid of the EPA, IRS and FBI—all in the name of IMS. (The federal government loves acronyms so we had all better get used to them.)

For those of you who have led pure and wholesome lives and don't know what about NIMS, it's the federally mandated National Incident Management System. NIMS was born in February 2002 when our supreme leader, George W. Bush, signed a presidential order directing someone in the federal government to devise an incident management system for the entire country. I love metaphors, so here goes: The president decides he wants to give every American 1 gallon of the finest applesauce ever produced. He calls his cabinet together and shares his grand applesauce vision, assigning responsibility to a newly appointed applesauce czar. The president tells the applesauce czar that he has one year to deliver a gallon of top-end applesauce in red, white and blue buckets to each American. Knowing nothing about applesauce, the applesauce czar directs his 500-person staff (who also know nothing about applesauce) to meet with every concrete company in the country to develop a formula for applesauce.

The presidential NIMS order has set incident command in the structural fire service back at least a decade; we are now rehashing debates that occurred more than 10 years ago. To create NIMS, the feds simply adopted the decades-old National Interagency Incident Management System (NIIMS), added a cover page and a lot of federal gobbledy-gook and renamed it the National Incident Management System (NIMS). The original NIIMS has become known as "two I'ed NIIMS" and federal NIMS is now lovingly referred to as "one I'ed NIMS."

Decades before President Bush mandated federal NIMS, there were two different IMS systems in American fire service: NIIMS (which was based on FIRESCOPE, the FIrefighting RESources of California Organized for Potential Emergencies) and Fire Command. The most notable difference between these two systems is that Fire Command refers to tactical subdivisions as "sectors," and NIIMS calls them "divisions" and "groups." In my mind, this whole NIMS thing isn't a matter of using either Fire Command or NIIMS, because the systems complement one another. This is where the feds really screwed things up.

Any moron can scream that something is broken, so in keeping with Fire Rescue's "Read it today, become an expert tomorrow, then land a high-ranking job with FEMA" philosophy, here is closure to the long running Fire Command-NIIMS debate and a simple solution for our federal IMS masters. NIIMS is truly an ICS. It was developed to bring together dozens of different fire departments, thousands of responders, and myriad other private, local, state and federal agencies at very large incidents, such as large wildfires. It was originally developed to include a huge logistical effort, manage a large aircraft component and track the costs associated with incident. NIIMS creates a temporary government to an area that has been devastated by some type of disaster, to manage long-standing incident control-and-recovery undertakings.

NIIMS is the logical backend model for NIMS. Despite what NIIMS advocates claim, NIIMS does not do the same thing that Fire Command does. Some may argue that NIIMS is scalable and all-risk, but it was never intended to serve as the day-to-day incident command system for fire departments any more than Fire Command was designed to manage a Category 5 hurricane that impacts seven states. Fire Command has evolved into a hazard-zone management system.

The intent and design of Fire Command's is to manage both service delivery (rescue, fire control and property conservation) and safety of the hazard-zone workers in real time. NIIMS incident accountability is performed on paper and tracks incident resources for cost recovery. Fire Command accountability allows the incident commander to know the position and function of all assigned resources. This type of worker accountability is designed around a 20-minute air supply in a burning building. Fifteen minutes is usually beyond the halfway point of a Fire Command operation. Fifteen minutes of a NIIMS event is merely a quarter of an hour.

We use Fire Command to manage day-to-day incidents (Types 4 & 5 as defined by NIMS) and anytime our members operate in or around a hazard zone. NIMS is used for large-scale, long-duration incidents (Types 1, 2 and 3 as defined by NIMS). These systems fit together quite nicely, as was proven during our department's two-week US&R deployment after Hurricane Katrina. The team managed their day-to-day rescue operations using our department's regular Fire Command system. Our team leaders reported to FEMA within the NIMS. This really wasn't that big a deal. Our team (just like all of the other teams) operated in very harsh conditions for a couple of weeks (conditions that killed alligators and other very hardy reptiles).

All of our team members came home without sustaining a single injury. They were able to pull this off because they operate within a structured hazard-zone management system every day. Our US&R team used the same hazard- zone management system they would use at a structure fire in Phoenix when operating in Louisiana, Oklahoma City and the World Trade Center.

In my mind, NIMS is the management system a Type 1 overhead team uses to manage the full scope of a big-deal incident. This includes managing all the various hazard-zone teams within that operation. Each hazard-zone team is managed by the set of bosses they showed up with. When our troops leave town to help someone else's hazard operations, we send a command team with them. Our command team will use our locally developed and refined hazard-zone management system to protect our workers. NIMS serves as the big system we all work within, and Fire Command is the system we use to tightly manage the task-level work and worker safety. Example: An overall incident includes 20 separate hazard-zone operations. A single, unified, all-encompassing grand Pooh-Bah command runs the entire incident while the 20 separate hazard zone operations are managed by their own command teams (or operations teams, semi-Pooh-Bahs, or whatever else the federal government wants to call them).

Early on in the NIMS process, the feds decided to deal directly with states, limiting the fed's number of customers to 50. This makes each state the "middle man" between the feds and every municipal fire department. The upside of this arrangement is the states and their cities have to figure out how to work together and manage Type 3 events, as well as Type 1s and 2s. On the downside, it places another layer of bureaucracy between cities and the ultimate NIMS authority.

The federal government has only been in the incident-command business for a few years. Based solely on their federal authority, they are now the experts. Many of the people providing advanced NIMS training have never managed an incident; some of them have never been to an incident. No one knew what to expect when this first got started, and initially, a lot of the federal NIMS representatives said the federal government didn't care what local fire departments did IMS-wise for local operations. Today, the feds are saying they still have no interest in local operations, but whatever we do on the local level must match up with what the feds do for Type 1 and 2 events. In the end, this may simply mean that NIIMS purists (wildland fire folks) have convinced the federal government to finally remove the word "sector" from the structural firefighting dictionary.

The federalization does not end there. In typical government fashion, NIMS really doesn't cause many incident responders to do anything differently, so there's really no need to fuss. For your state to receive federal highway funds, your local fire department, law-enforcement agency and public works departments must be NIMS compliant by a certain date. But what about fire departments (or other incident responders) that don't use an IMS? If a department gets all their members trained via the required NIMS classes, they become NIMS compliant. This same fire department can continue using an out-of-control approach for daily operations. Since most fire departments will never operate at the scene of anything larger than a Type 4 or 5 event, the federal government will consider that department NIMS compliant because their lack of a day-to-day system doesn't conflict with anything in NIMS. On the very remote possibility this same fire department does operate at a Type 1, 2 or 3 event, all they have to do is report to the right supervisor and run amok within the NIMS system. This creates a situation where organizations that have been effectively using IMS for decades must rework parts of their systems that don't need reworking, while organizations that don't currently use any type of management system can continue to deliver mediocre local service while exposing their members to unnecessary hazards.

Most Type 1 overhead teams are experts at managing large wildland fires. They do this better than anyone. The problem these current teams face is many of the incidents NIMS was created to manage are more urban in nature. These incidents include acts of terrorism in urban settings, natural disasters and epidemics/pandemics. Managing a burning forest does not qualify someone to manage a large explosion in the middle of a big city or figure out what to do when the people in your community start getting the bird flu. In the end, local responders will have to deal with these incidents well before federal help arrives.

Over the course of my career, I've noticed that when an incident begins well, it ends well. I have yet to read a firefighter fatality report that stated everything was going really well right up until the time Charlie got killed. This same truth applies to all incidents, regardless of size. Every year Florida gets pounded by hurricanes.

The country watches the news coverage, amazed by the wrath and power of nature. We also tip our hats to the effectiveness of Florida's emergency responders. This is a result of having local resources hooked together on the state level. The feds typically come in and support what is already in place and working.

Compare this to the aberration that was Hurricane Katrina. If the federal government is sincere about getting the American fire service to operate within a single system, we need another solution; NIMS falls woefully short. If the goal is to have all of us work together in times of extreme crisis, we need to work together on a day-to-day basis. Mandating each fire department to run automatic aid with each fire department on its borders would do more to unite the fire service in one week than NIMS has done in three years. If nothing else, it would be fun to watch.

Chapter 21
Frankenstein & Bambi

I recently had my yearly physical, and I'm convinced the staff of the Phoenix Fire Department's health center loves us. The examination takes a full morning and incorporates a lot of different processes, but the part of the physical I most look forward to is the lung-capacity test. It involves taking four deep breaths and exhaling as quickly as you can into a cardboard device that looks like a toilet-paper tube. It leaves you with a major buzz.

A few years ago while completing the pulmonary-function test, I experienced a brief yet complete loss of consciousness. The person conducting the test was a tall, slender female paramedic. She was on light duty while recovering from a job-related injury and was very excited to return to the real world of shift work the following week. Her voice and facial expression were saturated with enthusiasm as she described how the lung-test worked (she was a C-shifter). When she finished, I told her I would make her proud and proceeded to blow four of the biggest B-shift breaths ever breathed into a stupid cardboard tube.

I can't remember anything after the end of the fourth exhalation, but I woke to find our resident doctor and nurse pulling my large, semi-conscious ass off the bubbly C-shifter. I stood against the wall on wobbly legs, blood running into my eye from a very minor forehead laceration, as the medical team attended to Miss Perky. It doesn't take an A-Shifter to figure out what happens to a 120-lb. C-Shifter when a 265-lb. B-Shifter falls on her. She remained on light duty for another few months, but on the bright side I set the record for blowing into the pneumo tube.

We are fortunate that our department invested in a health center 20 years ago. Our department spends as much money maintaining firefighters as it spends maintaining its fleet of apparatus. This program keeps our members healthier—leading to better service delivery—and it gives our humanist fire chief a warm fuzzy every time he thinks about it.

For many of us, the health center serves as our primary care physician. Each year we receive a comprehensive physical. There are at least a dozen Phoenix firefighters still walking around today who would have died from some type of medical malady had it not been for their yearly exam. In many instances, our yearly physical forces us to take better care of ourselves because the health center's cardio-crazy staff has developed a comprehensive wellness and fitness program.

Part of our physical exam includes a blood analysis. A month prior to the company's physical, a health-center paramedic comes to the station and draws about 85 test tubes of blood from each member of the crew. There are very rigid rules for bloodletting, the biggie being that you can't eat for 12 hours beforehand. When we first fired up our health center, none of us knew about the 12-hour fast. Back in the early years, one of the health center's nurses came out to the station to draw blood. There were 16 of us, so it took more than an hour to bleed the group.

The blood technician had a specially designed wire basket that carried all the blood-laden test tubes in an upright position. By the time she had bled three-quarters of the group, we noticed a white substance that looked liked Sno Cap Lard floating on top of the samples. Before long, the top half of the blood samples appeared to be covered in shortening. When one of us pointed this out to the blood tech, she shook her head and asked when we last ate and what we had. We told her we finished a late breakfast of chorizo, egg and potato burritos right before she arrived. She gave us specific instructions about fasting prior to next shift's blood draw between bites of her own tasty, cholesterol-laden breakfast burro. The blood draw thing has triggered a memory I wish to share:

During the early years of my career, my brother and I toyed with the idea of taking the paramedic test. We both liked the idea of almost doubling our salary, and it was widely known that paramedics got to hook up with nurses, so it sounded like a fabulous career path. We signed up for all the required college classes and pursued our dream.

One of our classes was Anatomy and Physiology. Our teacher was a 30-something female doctor who smoked long, skinny, brown cigarettes. One day, my brother—a chronic and habitual tobacco chewer—told our very intelligent teacher, "You'd think that a doctor would know better than anyone about the causes of lung cancer." Our teacher took a long drag off her blunt, looked crossways at my brother, and with smoke wafting out of her mouth told him, "Medical science has come a long way. If I get lung cancer, they'll take a lung, but I'll recover and still look damn good. If your chewing tobacco gives you the big C, a doctor will cut off the bottom half of your face, and you'll end up spending the rest of your life drooling into a chin bucket."

One of the challenges with health and fitness is wading through the never-ending glut of nutrition information. Some experts claim bread is your enemy and mayonnaise is your friend. How in the hell do you eat mayo without bread? Others lecture that dairy products are the root of all evil. Remember when we were ordered to avoid eggs? The extremists advocate eating only raw, whole foods. What causes a person to delve to this level of insanity? Living on liquefied grass is fine if you are a goat or cow, but humans need a varied diet that includes carbohydrates, protein, fats and chocolates. Cooler heads are prevailing, and now butter and wine are back in vogue. We finally have something to thank the French for.

Nutrition is only part of the overall wellness equation. A few years before I joined the fire service, my department implemented a physical training (PT) program. At the training academy, we did PT every morning: stretching and calisthenics followed by a 5-mile run. I was never much of a runner, and my 5-mile jaunt usually took as much time as baking a 5-lb. meatloaf. One of my academy mates was a world-class runner, and he would complete the 5 miles twice, finishing up with me his second time around. My runner friend served as the physical fitness role model for our academy class right up until the time we started doing engine and ladder company operations.

He could run like a deer, but after 30 minutes of heavy physical labor, the bottom would drop out. One morning our class was doing hose lays; it was our fourth or fifth evolution, I was the plugman and my runner mate was the nozzleman. The engine stopped at the plug, our captain yelled, "Take the plug," and I jumped off the rig and proceeded to take the plug. I finished with my plug work, jogged up to the rig, put the hydrant wrench back on the truck and donned my SCBA. As I came around the side of the apparatus, I saw the highly aerobic nozzleman struggling to pull the 1.5" hose load out of the transverse hose bed. One of our instructors studied a stopwatch intently as he screamed, "What are you doing? There are children burning to death while you're fucking around Dorothy! You've got 45 seconds left to pass this hose evolution!" If we flunked, we ran. I hated running, as did most of my other classmates. It was evident the gazelle was not going to get the line into operation, despite the instructors' motivational lectures and the fledgling firefighters pleadings. I walked over and grabbed my struggling classmate, put my arm through the hose load loops, picked him up and pulled the hose bed. I moved at the same speed as I did during my morning run, only this time I was carrying 170 lbs. of fully turned out, top-flight aerobic firefighter and a charged attack line. As our instructor told us we passed our hose evolution (and wouldn't have to run laps), my little runner buddy puked and passed out.

The fitness level and overall wellness of our members is what our health center is all about. The end goal is to have a healthy and fit workforce. This broad goal allows our healthcenter staff to focus on each individual member, making recommendations that will improve both the quantity of life and quality of health. To a lesser degree, these health-center recommendations have the potential to improve each member's physical capacity to deliver service (i.e. drag hose lines, beat their way into burning buildings, lift really heavy customers, etc.). Off to the side is another large issue of what type of physical attributes do we use to test for the position of firefighter? This is where what looks good on paper collides with reality.

Most doctors, exercise physiologists and other medical professionals would have pronounced my runner classmate to be the fittest person in my academy class—in fact some of these people did. The problem is firefighting has nothing in common with running, rowing, ellipticalizing or any other aerobic activity. On the other hand, when we do these activities as part of a regular fitness routine, we get in better shape, which can result in higher job performance. But firefighting still remains a very anaerobic activity.

The challenge for all of our organizations is to hire people who can do the job. Over the last decade our profession has refined the physical entrance exam for the firefighter's position with the CPAT (candidate physical activity test). The CPAT was a joint effort between the IAFF, the IAFC and 10 different fire departments. Phoenix was one of the cities that participated in this process. As one would imagine, a lot of different groups were involved in developing this test, including doctors, exercise physiologists, fitness trainers and various A-shifters. One of these early development teams was assembled at the training academy in my own hometown. One of the early CPAT courses was laid out on the drill grounds (or as we call it, The Grinder), and a group of guinea pig firefighters ran through the course. Most of these lab rats represented the best and brightest our department had to offer. The gathered group of experts was thrilled beyond words as they monitored Phoenix Fire Department's super humans. The test was going as planned when one of our training-academy staff asked if they could send one of our department's future recruits through the CPAT course. The clinicians welcomed the opportunity to test one our future bionic firefighters.

One of our training captains went off to retrieve Aaron, who had been welding props for our burn house all day. He came out wearing a welder's apron and helmet with Frankenstein booties adorning his canoe-sized feet. He stood six and a half feet tall and weighed more than 300 pounds. He was the size of an entire ladder crew. The out-of-town clinicians looked at him like he was some kind of new virus. I don't blame them. He was a lot different than the Aryans our fitness Nazis had loaded the test with. Aaron finished the CPAT with one of the day's best scores, but that's not what had the gathered group concerned—it was the way he went through the course. It was almost too easy for him, and it looked like he was going to break all the props. One of the observers commented that when Aaron picked up the dummy for the rescue drag, it was so terrified that it voided the contents of its bladder.

Some of the clinicians voiced concerns that a man of that size would break down over time and eventually become a casualty of his career. They thought smaller, more aerobically tuned humans were better suited for a career in the fire service. This became such a big issue that the group took it to our fire chief, who patiently listened to the fitness scientists. When they were done voicing their concerns he told them, "If I had to pick any one person involved in this test to go into a burning building with me, I would choose Aaron." Aaron's career is coming up on 10 years, and he's a more capable firefighter today than he was on the day he confused the group of CPAT experts. My gazelle-like classmate ended up washing out of our department after five years. Go figure.

Humans are driven by their basic biological urges. To support my hypothesis, I point to our closest mammalian relative—the chimpanzee. Human and chimpanzee DNA is more than 98 percent identical. The biggest differences between man and monkey are that humans string together words in verbal and written forms and chimps don't fire other chimps for viewing perfectly legal porn at work.

Chapter 22
The Lust of Monkeys

Every year hundreds of firefighters lose their jobs over perceived misbehavior. It could be argued that being fired carries the same professional effect as dying in the line of duty. The member may not have died, but as far as the department is concerned, they no longer exist. It is not my intent to minimize dying in the line of duty—only to expand the definition and shed some light on possible strategies that will allow all of us to enjoy a long retirement.

Some of you may be thinking that I have finally gone over the edge, but before you jump to a hasty conclusion, think about this: Imagine you are at home and you've consumed too much hard liquor. Your gaze wanders over to your next-door neighbor's house, and you notice a soot-faced lamb grazing in the front yard. Before you know it, the chimp region of your brain takes over. Your interspecies love tryst is interrupted by both the homeowner and local law-enforcement personnel. (*Note:* This is a true story.) I would rate the consequences associated with man-sheep love to be on par with having a wall fall on you. In some ways, it is even worse. When a wall falls on you, you die a hero's death; your family doesn't leave you, and you don't have to change your name and move to the Arizona-Colorado border because you will forever be known as the guy that nailed the next-door neighbor's farm pet.

The newspaper headlines are routinely littered with similar tales regarding our inflamed passions, so in keeping with this magazine's tagline, "Read it today, do it doggy-style tomorrow," let's get this party started. Within the last several weeks, the festivities at a fire department award banquet shifted into high gear when several of the rank-and-file attendees allegedly engaged in sex on the buffet table. If this tale is true, it is every bit as tremendous as it is inappropriate.

As if it weren't hard enough to control one's sexual appetites while on duty and at formal department functions, our workforce finds ways to compound the problem, as the following examples demonstrate. Many departments use intranets to facilitate training and transmit messages from the grand and exalted chief of the department. Some of these computers include Web cams. As the story goes, a firefighter was conversing in a chat room with someone he believed to be one of those hot, young girls gone wild. The online conversation turned racy, and soon our doomed firefighter was coerced into manipulating his little nozzleman for the mystery woman on the other side of the Web cam. It doesn't take a leap of imagination to figure out the mystery woman was actually five of our doomed firefighter's buddies in various stations around the city. Since this act was witnessed by more than one firefighter, word quickly spread, a formal investigation ensued and everyone involved was immediately fired. Payroll stopped paying the terminated members as surely as if they had been killed in the line of duty.

Internet porn continues to be a huge distraction for organizations and has become the baseline misdemeanor most departments struggle with daily. Many fire departments have established zero-tolerance policies for workers who look at sexually explicit material on the company's dime. Fifteen years ago, my department established a policy that prohibited posters of scantily clad women hanging in public places within the fire station. It was explained to the workforce that the department really didn't have a problem with posters of scantily clad women, but it did have a problem with the nuisance of processing complaints stemming from posters of scantily clad women, so would you please take them down and save all of us the time and nonsense? Even us B-Shifters shrugged our shoulders and took down the posters. To our department's credit, if an offensive poster was found in a public place, it was quietly removed and no one was suspended, demoted or executed.

Our society expends lots of energy establishing standards for what it deems acceptable. One end of the spectrum is occupied by the moral puritans, easily identified by their high collars and long noses. The other sideline is occupied by a group of Eastern European pornographers, high on a never-ending supply of money and shit-stained fists. These two combatants come equipped with a bottomless pit of moral edicts, attorneys, lobbyists and Supreme Court decisions. The rest of us are left to navigate the minefield they have created in the center of our professional lives. No employer in their right mind wants to end up in a wrestling match with these lunatics, so we establish zero-tolerance policies and hope the workforce obeys. Life is never this simple.

185

Before we all crawl into our fire department bunkers and proclaim termination as the punishment for a low-impact first offense, answer a simple question. Which activity would most normal, clear-headed people define as being crazier: 1) viewing a pair of perfectly developed coeds giving one another a firm but loving spanking or 2) running into a burning building? We hire young, aggressive, pretty, hormone-saturated firefighters, and we shouldn't be shocked when one of them has a biological chimpanzee moment. I am not suggesting that we turn fire stations into Roman-style vomitories. The workforce shouldn't be watching "touchy-feely" while on duty, but the consequences should fit the offense. Most Internet pornography is not illegal. If it were, overzealous prosecutors would have thrown many of the residents of California's San Fernando Valley in jail (this is where the majority of American porn is produced). Therefore, it can be argued that viewing pornography in your office on your work computer falls into the same category as booking a personal vacation, shopping for the latest MP3 player or applying for a hunting permit.

It may make us feel all-powerful to implement and enforce a set of rules for the organization, but we are not the final authority. There are a multitude of organizations that review the final outcome of our personnel processes. These groups include the local authority we work for, the media, the community, everyone's mom and the entire judicial system. Within the last few months, six firefighters who had been terminated for on-duty hanky-panky had their jobs reinstated. Imagine terminating of one of your members for viewing garden-variety Internet porn, then having to process it through a court system that refused to prosecute a beautiful 25-year-old female teacher who gave extra credit blow jobs to a 14-year-old student. Good luck.

If your head doesn't hurt by now, consider the case of a Tennessee fire department member who sued over the use of a take-home car and inaccurate pronouns. The member used to be an alleged male, recently transgendering into a female. In this case, the flesh wasn't weak, it was simply wrong. After returning to work, the member lost her take-home car and her co-workers began calling her "sir." If I showed up to work one day as Dominique Brunacini, and all I had to deal with was not driving my response vehicle home at night and having the other shift commanders call me "sir," I would be all right with that.

When I began my career, many large metropolitan fire departments were operating under judicial consent decrees. These federal court orders pertained to hiring practices along with cases of workplace harassment. The advancements in the fire service don't have much to do with eliminating these complaints because many departments still have judges managing their personnel sections. Where we've really made hay is in the diversity of our members who continue to get themselves into hot water for what was once the domain of high-ranking white males. It made my heart soar like eagle to read about the fire chief who has several sexual harassment complaints pending. In the good old days, it was the exclusive territory of the men to harass the women after we grew bored with harassing one another. Welcome to the 3rd millennium. The chief involved in this particular case is a woman. She is accused of mismanaging her personal relationships with some of the female members of her department. In the process, she has exploded the glass ceiling that once kept females out of positions where they could potentially abuse their power and authority.

We males have hoodwinked ourselves into the fantasy that we are the masters of the universe. One only needs to look to nature to find out that the female does most of the hunting and heavy lifting, and bites the head off her companion when she is finished mating. Where does this leave us? For years our personnel sections have hounded us to hire a workforce that reflects community. Mission accomplished. I don't think we need to develop a whole new set of complicated rules to deal with today's workforce. What works the best is to hire good people, train them, let them know the organizational expectations and treat them the way you want them to treat you. This simple philosophy works equally well for men, women, minorities, transgenders and B-shifters.

Because we hire human beings (who sometimes act like chimpanzees), they will make mistakes. When you make a mistake, how does it make you feel when your boss or department's first reaction is to fire you? The other side of this coin is if the members engage in behaviors that can lead to a prison sentence, they shouldn't be surprised if the department lets them go.

I work in an automatic aide consortium of 25

departments. During the last 10 years the organizations in this system have lost approximately 20 of our members as a result of bad lifestyle choices. These lifestyle decisions typically involve the things they put into their bodies, putting inappropriate parts of their bodies into others while on duty and a singular off-duty act of man-sheep love. If we had two line-of-duty deaths a year for the past 10 years, every safety organization in the country would be crawling up our skirt. In the meantime, substance abuse draws a yawn, empowerment and a 30-day treatment program, while sexcapades get a snicker-filled week of headlines and local newscasts, followed up with stern lectures and strong discipline.

I do not mean to digress but I would like to make an announcement—the national Everyone Goes Home safety initiative really doesn't go far enough. This program, which the National Fallen Firefighters Foundation launched in 2004, is an excellent effort designed to eliminate firefighter fatalities and injuries. Every year, more than 100 of us are killed in the line of duty, with tens of thousands more suffering serious injury. I have two very simple solutions to these problems. Part one deals with reducing vehicular related fatalities: We remove all the lights and sirens from our apparatus and respond Code 2. The second part will virtually eliminate all tactical fireground fatalities: We no longer conduct offensive, interior operations. I am fully aware that these suggestions fly right in the face of 300-plus years of tradition, and they don't stand a chance of ever being implemented.

Some may also argue that they probably don't line up

with our mission. Maybe not, but I don't have enough room left in this edition to jump down that rabbit hole, so let me end with this thought. It is our personal responsibility to manage our day-to-day professional lives in a way that leads to a long and harmonious retirement. It doesn't matter if your career ended short because a burning building killed you or you got caught banging a 16-year-old girl in the station hose tower. No one will care that she looked like she was 23 and fawned all over you because you looked so sexy in that wool coat.

Chapter 23
Success Makes You Stupid

Every month our fire chief has a series of meetings with all of our department's chief officers. These meetings are repeated over three days, so all of the response chiefs working a 56-hour week can attend while they are on duty. This is really nice because a day off really isn't a day off if you have to come in for a meeting. Our latest gathering included a presentation from John Tippet, the project manager for the National Firefighter Near-Miss Reporting System. John is a battalion chief with the Montgomery County (MD) Fire Department; he is an excellent presenter and gave the group a brief yet comprehensive overview of the program. Their Web site is designed for firefighters to share their near-miss experiences in an effort to keep similar situations from injuring or killing others. The program is modeled after the FAA□ s successful near miss reporting system, which has drastically improved airline industry safety.

Some of the best training my department has ever done is "near hit" incident reviews. Each video segment is based on real-life incidents where one or more of our members narrowly escaped serious injury or death. Our department's safety section developed this program. The safety officer gathers any available video footage of the event (sometimes even recreating the near-hit), conducts on-camera interviews with all of the incident players and distills the lessons learned from what happened. The training is delivered in the stations by the safety officers.

Sharing lessons learned from real events, as told by the people who actually lived through them, goes a long way in breaking through the "it can't happen here" mentality. In this spirit, I offer up the following experience as my effort to help seed the National Firefighter Near-Miss Reporting System so we can all retire and take more out of our pension systems than we put in.

This tale involves an entire station I used to be assigned to. I was the captain of a ladder company that went by the name of the "Raging Queen." We ran out of a station that also housed an engine company and a rescue ambulance. One afternoon between the hours that separate lunch from dinner, we were dispatched to a reported structure fire. The address was a half-mile south of our station. The dispatch ended a scolding we were receiving from our district commander. The C-Shifters had contacted our chief to complain about all the signs that decorated our station's bathroom. Over a period of years, the crews had hung road signs on the bathroom walls. There were all kinds of traffic control signs, street signs, "Welcome to the Grand Canyon" signs, as well as a sign proclaiming: "Adobe Mountain Correctional Facility Ahead. Do Not Stop For Hitchhikers." Every square inch of empty wall had been covered with roadway art. A week or two earlier, our chief showed up and ordered us to take all of the signs down. He felt there was a possibility the city could be charged with possessing stolen property. One of the crew members asked our fearless leader what we should do with all the signs once we got them off the walls. He told us to take them home. He wasn't a smart man.

Holding to true B-shift form, we felt we could take bathroom decoration to a new high without exposing the city to any further legal risk. After taking all of the signs down, we patched the holes and painted the bathroom City of Phoenix beige (this is an approved color). The next morning, C-Shift was absolutely glowing over the plain Jane, vanilla restroom. It turned our stomachs as the group of them denied complaining to anyone who would listen to their incessant sniveling about our sense of style. We simply smiled and left, knowing full well that next shift we would raise the bar.

One of the things C-Shift disliked about us was that B Shift had hired a mildly retarded, Native American as a houseboy. To fully describe Mr. Blueberry would require a full chapter in a really big book. Suffice it to say that Mr. Blueberry's head looked liked a large Halloween Jack-o-lantern on very strong LSD. We took a close-up photo of Mr. Blueberry smiling his patented crooked smile. Just like the taste of chocolate, mere words cannot describe the full effect of the picture.

We had dozens of 8x10s printed, and we wallpapered a border around the top of the bathroom. We had a few pictures left over, and glued them over the centers of all the clock faces after captioning them with, "Mr. Blueberry says it's time to quit slacking and get back to work." It was eerie the way Mr. Blueberry's eyes watched you as you urinated or followed you around any room that had a clock in it. Needless to say, the C-Shifters had an adverse and hysterical reaction. True to form, they called the chief to file a new complaint. Our chief had showed up to the station that morning and was laying down the law when the voices came out of the wall and told us to go to the burning building a half mile south. As were headed out to the rigs, he warned us that we hadn't heard the last of this.

Both the engine and ladder companies were in quarters. The group of us turned out, mounted our steeds and gunned the motors as we responded toward our uncertain future. As we turned onto the street, I looked in the side mirror and saw our chief standing in the exhaust-filled bay, moving his mouth and shaking his head. I looked up and saw our engine company make a hard left down the street that bordered our station and head south. Our rescue ambulance had been on their way back to the station when the call came in, and it was heading toward us. Not wanting to wait while the engine and ladder were turning down the street that would take us to the fire, they opted to turn down the alley that ran the length of the block. It appeared to me that the ambo driver was fully turned-out, and his attendant partner was attempting to don his bottle in the tight confines of the ambo's cab.

As I turned to look south, I saw most of the sky blanketed with a thick carpet of black smoke. The volume and intensity of smoke and fire sent me into my special reflective place. This private time was violated when Rodney the engineer said, "Look at our idiot ambo crew." I turned my head and watched as the ambo paralleled our same course—only we were driving down a paved road and they were attempting to navigate their box down an alley. The driver was having trouble steering because he was wearing his SCBA and had his chest smashed into the steering wheel. The ambo was trying to pace the ladder. The only evidence of their journey down the alley was a rooster tail of dust. Every so often, the landscape cleared enough between low-rent houses to actually see the chaos.

Midway down the alley, the ambo clipped a large black plastic trash container that sat in front of a chain-link fence. The force of the impact drove the large trash bin into the fence, turning it into a giant slingshot. As the ambo flew past, the trash container was launched across the narrow alley, knocking a large chunk out of the neighbor's concrete wall. A few seconds later, our ladder truck made a right onto the main drag and headed toward the fire. The alleyway that our ambo crew mistook for a boulevard dumped out onto our road. The long, narrow alley looked liked it had just hosted a tornado. All manner of trash and debris were floating in the half-mile cloud of dust. As we drove past the ambo, we could see that both firefighters were fully turned out, with their SCBA masks on. I felt like I was trapped in a Jules Verne novel.

The scene that lay before me was quite impressive. An old, stand-alone tire business was vaporizing in spectacular fashion. The owners had relocated their operation and were in the process of moving out. Most of their equipment had already been removed, leaving only old, used tires in the main part of the building. The peaked roof sported a large hole in its center with a massive volume of fire blasting its way toward the heavens. The entire sky was blotted by thick, black smoke, laden with heavy carbon particles. The little sunlight that cut through the dark sky hit the smoke and sparkled, making it appear that thousands of tiny little jewels were flying through the air, while large chunks of ash rained down around us. Between the noise, power, heat and visual spectacle, it felt like we were standing next to a volcano. It was one of the rare occurrences where people were running away from the scene.

Our engine had laid a supply line. We parked behind them and both of the truck engineers began setting the rig up for a defensive operation. In the meantime, the ambo parked and both the firefighters joined us in front of the building. Our plan involved taking a couple of attack lines to the rear of the burning building to protect the houses and businesses that bordered the north and east sides of the fire while the engine and ladder used master streams on the main body of the fire.

The building was surrounded by a 6' chain-link fence topped with three strands of barbed wire designed to keep neighborhood prostitutes away from the tires. Six of us stood outside the main gate when the firefighter off the ladder grabbed the bottom strand of barbed wire and pulled himself up and over the 7½' high barrier. This caused several of the firefighters on the wrong side of the fence great agitation, because they did not possess the physical attributes of a world-class gymnast. What they lacked for in scaling ability they made up for in sheer power - having the strength of large farm animals— which they used to hurl themselves into the fence. This had the same effect as hitting a large trash container at 45 mph with a three-ton ambulance. They moved the entire section of fence 4 to 5 feet, and it catapulted two of them through the air for a short distance, then onto the ground. Eventually, we cut the chain on the gate and the group of us caught up with our nimble truck-company firefighter.

We determined there weren't any critical exposures at the same moment both of our attack lines were charged. The back of the building had lots of wide-open doorways that provided us with good access into the burning tire shop. There wasn't very much of the building's interior space that wasn't burning. The building didn't have a ceiling, and we could see all of the wood framing that made up the roof burning merrily. At the time, it made all the sense in the world to radio to our guys in the front and tell them to hold off on the outside master streams for a minute or two while we tried to knock it down from the inside with handlines.

All the heat was going straight up through the ever-increasing hole in the middle of the roof. Most of the tires had been eaten up by the time we got our two lines on the interior. It took us a little while to knock down the fire below, but we spent the majority of our time extinguishing all the fire above our heads. It was the best 15 minutes that any human spent on planet Earth that day. We started running out of air around the same time we had the fire knocked down. Exiting the building, the group of us stood together in the back all by ourselves. No one really said much; we were still lost in the high and in mild disbelief over the fact that we did what we did. We had been out of the building for no more than two minutes when we were pulled from our moment of reflection as the entire roof suddenly collapsed with frightening energy, causing half of the rear wall we were standing in front of to do the same.

We spent the next half-hour wrapping things up. During this time, other firefighters, company officers, and chiefs congratulated us for our miraculous stop. Even Chief "Stop Decorating Your Station" took the time to tell us how fabulously we had done. None of this changed the fact that the group of us had been a handful of seconds from wearing many tons of absolutely worthless property. I am fairly confident that we would have all been severely killed had we been under the roof when gravity had its way. Sometimes success makes you stupid. This lesson was punctuated a couple days later when the local demolition company hauled the remains of our tactical triumph to the dump.

The challenge we face is to somehow make sense of an activity that fills us up on a very personal level. Structural firefighting is its own reward. It is the reason the opposite sex and the mentally retarded are so drawn to us—we will voluntarily run into burnings buildings without hesitation or a second thought. Sometimes we disguise our insanity with gift-wrap and a pretty bow, justifying our actions with, "We are the public's last chance at survival," or any other catchy phrases that enables us to engage in high-risk/no-gain endeavors.

Maybe we should frame the concept of risk

management in a different light. Let's assume that you have just purchased a brand new Cadillac. You have just driven it off the lot and suddenly notice an old, dilapidated tire store well involved in fire. In the back of your mind, you know the building is toast and anyone unlucky enough to still be inside is deader than 10 mofos. Would you drive your new Cadillac into the building? Of course you wouldn't. You wouldn't drive any vehicle you own into a burning building. That would be insane.

Chapter 24
Kool-Aid Horror

Christmas is right around the corner. If this year brings the same "what are we supposed to call the holiday" confusion as last year, let me make a suggestion: Quit whining about the big retailers renaming "Christmas" to "Holiday" and get over it. These companies are in business to sell you stuff, and they'll eliminate anything they perceive as getting in the way of the bottom line. In today's world, nothing is more divisive than all things religious and spiritual. The only thing that would cause these retailers to once again refer to the celebration of Jesus' birthday as Christmas is if you mackerel snappers quit buying stuff from them, in favor of spending your money at Jesus-friendly stores. The real rub here is the "Holiday" stores sell their Chinese-made goods cheaper than the "Christmas" stores sell their Rangoon-made gift bobbles. The whole thing makes me want to go get an abortion.

As many of you may have gathered, the topic of this month's article is the politically correct homogenization overtaking us. The vast majority of people simply want to live and let live. This runs counter to the people we are bombarded with each and every minute of every hour in the media. To be heard in today's world, one must be a zealot, a criminal or an idiot. It doesn't matter which side of an issue you take, as long as it is fundamentally fanatical and in direct opposition to the group of lunatics living on the polar opposite side. As a result, much of center mass of our civilization is forced to hide behind the skirt of political correctness in an attempt to avoid drawing the wrath of the xenophobes currently in charge.

The driving force of many bosses, managers, high-ranking clerks and community leaders has become, "How will the public react?" But how do you define "the public"? All too often, the public is mistaken as the vocal and overrepresented minority of lunatics who have nothing better to do with their lives than run in an ever-diminishing circle and react. An excellent example: the reoccurring issue in my own community about nursing mothers. Every three or four years, some hypersensitive person becomes enraged when they see a baby with a mouthful of nipple, and the rest of us get to read about it in the paper and listen to the evening newsreaders talk about the societal ramifications of lactating mothers. Politicians are pulled into these types of issues like a drunk driver is drawn to oncoming headlights. Before you know it, we are legislating for or against the simple biological need of motherhood. This becomes criminally ridiculous when you consider that somewhere in this solar system is a comet the size of Alaska that has the Earth's name on it. When it hits our fluffy blue planet at 30 miles-per-second, nursing mothers and most other mammals will become roach chow. Let me go on record as being offended by planet-hitting comets.

Before I go much further with this, I'm going to have to relate it back to the fire service. The magazine has been very insistent about this, so in keeping with the theme of this fine trade journal—"Read it today, file an injunction against it tomorrow"—I will turn the mirror of internal reflection on myself and the rest of the fire service.

Fire department families are just like your regular everyday families—some are nurturing and healthy, while others are dysfunctional and caustic. We also reflect the quilt of American society. Members of fire families have formed clubs, organizations, associations and other splinter groups that toss back the vanilla-ization of our "craft and trade." The Christians, Italians, Jews, Amish, volunteers, Mexicans, lesbians, gays, women, cowboys, Irish, skirt-wearing bagpipers, hunters, power-boaters and card players are just a few of the groups that have formed alliances to pursue their common pastimes and advance whatever agenda they feel needs advancing. Many of these groups formed in an effort to protect themselves from some other part of their fire department family.

Within the last six months, I spoke with an intelligent, capable, high-ranking member of a good-sized fire department. His department just finished the company officer promotional exam, and a large number of women ended up at the top of the list. He lamented: "I can see promoting one or two of them, but not 40 percent of the new officers over the next few years." What makes this story really come to life is this individual belongs to his own splinter group, and as few as 20 years ago it would have been very difficult for him to be promoted due to his ethnicity. Ain't life just crazy? I suppose intolerance is a family-learned trait that doesn't discriminate based on skin color, gender or sexual preference.

Most of us go out of our way to get along with others. The reason for this is pretty simple—you've got to go along to get along. When we allow people their differences, they allow us ours. This is the glue that holds society together. In the close-knit quarters of fire departments, the leeway we allow one another becomes the ultimate factor in how good a place it is to work. Fire departments that strive to make everyone the same end up with very large volumes of rules and regulations that pertain to personal conduct. These same places always come equipped with large personnel divisions that are kept busy with investigations, enforcement and dealing with the avalanche of lawsuits that stems from trying to make everyone the same.

I am not advocating total anarchy, merely suggesting that the few rules we do maintain be developed around what we do for a living, with some of the U.S. Constitution thrown in for good measure. A rule that says we will come to a complete stop at a negative right away makes a lot of sense. We are a public safety response agency and shouldn't kill the motoring public as part of our standard response. On the other hand, 100 pages of rules that apply to uniforms and personal grooming standards would be considered a tad bit obsessive by most mental health clinicians.

As a culture, we spend an inordinate amount of our time and energy being offended by inconsequential bullshit. Recently, a kitty-related event occurred at a fire station. One of the shifts at the undisclosed fire station wanted lemonade stocked in the kitty. The kitty man acknowledged the request and agreed to purchase the refreshing lemon beverage on his next kitty-shopping safari. Over the next several shifts, the kitty man grew weary of the requesting shifts' constant pleas of, "When are you going to get our lemonade?" So one morning he made a special trip to buy a big bucket of powdered lemonade drink mix. He was deeply disappointed to discover the store had recently sold their last pallet load of fake lemonade. When the kitty man explained this to the lemon-lover shift, it did not sit well with them. Their disparaging comments about his kitty-management skills were mean and unnecessary.

The next shift, our kitty man and his crew were shopping for that day's two tasty meals when he remembered the lemonade. Searching the store shelves, he found Kool-Aid brand drink packets and threw a dozen of the lemon-flavored packets into the cart. When he got back to the station, he placed the Kool-Aid into the requesting shift's kitty locker. The next day, the lemonade lovers went to the station kitty locker (as opposed to their shift's kitty locker) in search of a drink mix that would pucker their pie holes. Failing to find any lemonade, they voiced their frustration with the oncoming shift about the incompetence of their station kitty man. This caused a turn of events, because the lemonade-loving shift and their relief shift were generally at odds with one another over every single issue in the known universe.

The off-going shift told the lemon lovers to, "Quit being a bunch of whining cunts and get a life," and then wished them a fabulous shift. The next shift cycle, the kitty man reported to duty and was given a full accounting from the off-going shift. The kitty man replied, "They are morons. I bought them their damn lemonade over a week ago and put it in their shift kitty locker. I'll take care of this right now." The kitty man retrieved the small, brightly colored packets and taped them to the door of the lemon-lovers' kitty locker. Another full rotation of the earth passed, and the lemon lovers were back on duty. Sometime after relieving the kitty man's shift, they found a dozen packages of lemonade-flavored Kool-Aid taped to their kitty locker.

Instead of being overjoyed with the prospect of a shift filled with lemony goodness, they called their battalion chief to inquire about the feasibility of filing a formal complaint against the shift that normally relieves them. As luck would have it, many of the members of the lemonade shift belong to an ethnic group that has stereotypical ties to Kool-Aid products. The lemon-lover shift assumed the other shift was sending them a racially motivated jab. Nothing could have been further from the truth.

The indignant lemon-lovers escorted their battalion chief to the kitty locker so he could see the offensive act for himself. I really don't know what private thoughts the BC had as he stared at a cabinet door with a dozen yellow packets of drink mix taped to it. Perhaps he was thinking to himself, "My God! The inhumanity of man against man," or maybe it was more along the lines of, "You've got to be kidding me. These morons called me for this?" Whatever it was, he kept it to himself, but he wasn't taking any chances. He passed it up the chain to his boss. This resulted in a station get-together the next shift between the lemon round-up gang and the shift they were accusing of drink-mix racism.

The accused shift was completely unprepared for what awaited them. That morning they were greeted by an assortment of chiefs and a group of lemon-lovers loaded for bear. One of the attending chiefs began the proceedings: "We are here to determine what your intent was in taping packets of Kool-Aid to such-and-such's kitty locker." The oncoming shift shared the curious and confused facial expression of a dog being summoned by a dog whistle. Almost in unison they replied, "What in the hell are you talking about?" The lemon-lovers were full of moral outrage and they bellowed, "You knew what you were doing when you taped that to our kitty locker, and don't try to deny that you are messing with us. This morning we found a watermelon in the refrigerator. Was that just an innocent accident also?"

Now most the people in the room were looking at one another like they had just seen an enchanted elf riding a flying giraffe. The accused shift was having none of it. One of them came forward and told the assembled throng, "You're out of your minds. Most of us weren't born in the big city. We weren't privy to the fact that one ethnic group favors Kool-Aid more than another. We didn't tape it to your locker, either. As far as the watermelon goes, keep your hands off of it. We bought it for dessert last shift but didn't get around to eating it because we were too busy. You guys really need a hobby."

During this matter-of-fact lecture, the kitty man magically appeared and asked, "What's up with the meeting?" The lemon lover spokesman cried, "These bastards taped Kool-Aid to our kitty locker and now they're trying to deny it." The kitty man broke into laughter and said, "You guys are a bunch of diva bitches. I did that. You asked for lemonade and sniveled and whined and bitched every shift you had to go without it. I bought it and threw it in your locker over a week ago and you idiots couldn't find it and cried even louder. I taped it up so you couldn't miss it." This effectively brought the inquisition to a screeching halt. Once again, a kitty man had led them out of darkness.

Chapter 25
My Sweet Ride

One of the joys of my life is traveling with my dad and getting to hang out with firefighters all across this fabulous country of ours (and parts of Canada, too!). The fire chief is still very innocent when it comes to any electronic device that doesn't reshape wood and steel or compress air, so my brother or I will often tag along on his teaching adventures to make images appear out of computers. A few months ago, the two of us presented a leadership program on the West Coast. The only downside of our modern travel adventures is the few bad apples who work for the TSA; they can easily set me off. On more than one occasion, I have been so traumatized by the behind the screen cock-and-ball fondling that I've blurted out derogatory statements during some of our out-of-town training sessions. During my latest tirade, I told an A-shift paramedic that his small elbow and knee joints made his arm and leg musculature appear much larger than it actually is. I wish to take this moment to apologize for these tasteless comments. In hindsight, the IAFC may have been right when they called me a "callous and insensitive Neanderthal who must be stopped."

During our recent leadership fiesta, I was sitting off to the side in rapt amazement while Big Al carried on with the crowd of highly enlightened fire-service professionals. A group of 75 firefighters were engaged in an active customer-service dialogue. The fire chief was making the point that all leadership must be based on our core mission. He gently took all of our brains by their little hands and led us on a journey. "I've read hundreds of books on leadership. I left most of them on airplanes after struggling through the first 100 pages. Maybe the people who clean airplanes can make more sense out of them than I could. One day it occurred to me that all this leadership material doesn't become relevant, or make a bit of sense, until you can attach it to something tangible. The only tangible thing we do in the fire service is deliver service to our customers. When I finally figured out everything must be based on the customer-service relationship, it framed the entire role of leadership for me."

More than a dozen people clamored to all talk at the same time. Positive energy, love and rainbows were exploding throughout the auditorium. "We are nothing without our customers!" "There isn't anything we wouldn't do for our customers!" "We must rewrite all of our procedures around five-star customer service!" "All of our management practices must address how they effect customers service!"

It was a real dawning of the Age of Aquarius moment. The group was very enlightened (to say the least). I sat and observed without saying much as the group ignited and glowed, high on the hope of a better tomorrow. I couldn't help myself anymore and floated a turd into the punchbowl. "All this new-age crap is really sweet, but how do you manage and lead a subordinate who just doesn't buy into all of this?" This caused an awkward silence for a second, then the crowd reacted like angry protestors. I remember an attractive, athletic, blonde, female officer responding: "We simply should not hire people who do not buy into the concept of high-quality customer service. They have no place in today's fire service." This was followed with a dozen like-minded people scolding me for suggesting that a member of the fire service might not buy into the concept of quality customer service.

I have the innate ability to appear that I am following along with a conversation when actually my mind is somewhere else. While the crowd of hummingbirds attacked me, I floated 10 years back in time. All the love, leadership and serving our fellow man talk transported me to an event that would have frightened and incapacitated most of the people in the room. I have had buildings fall down around me, seen people cut in half and have had bosses threaten my employment, and none of these experiences tested me like this singular event. It opened my eyes to the very nature and depravity of humans like nothing before or since. The true nature of leadership lay at the core of this phenomenon.

My wife and I have three daughters. The first two daughters were normal little human girls. Our final daughter looked like a regular baby girl. After spending the briefest period of time with this beautiful child, it became apparent that the soul of a Viking warlord was trapped inside her. One afternoon grandma came by to take our Little Angel out for a day of female bonding. When it came time for granny and her granddaughter to go, the Sven the Destroyer inside her decided she wasn't going anywhere. Granny made the mistake of believing that a mature and experienced woman could make a small child fall in and march straight. I made the suggestion that granny shouldn't tempt fate and save the day out for another time. This earned me a look that said, "I've been dealing with kids a lot longer than you, Mr. Stupid Man. A 1½-year-old child will not set any agenda of mine." She had the good manners to simply tell me, "We'll be fine." I smiled and said, "Have fun."

Granny picked up the child and got as far as the front door before the first signs of demonic possession set in. Writhing of epileptic proportions and speaking in tongues did not scare nor deter granny. The old woman gripped her granddaughter tighter, gritted her teeth and pressed on. The battling duo had made it halfway to the parking lot when my Little Angel positioned herself where she could kick her grandmother right between the eyes. This knocked one of the lenses out of granny's sunglasses, took both combatants to the ground and allowed our Sweet Cherub a temporary escape. Granny quickly shook it off, rounded up her little charge, stuffed her in the vehicle and went on with their day. During this period of our lives, we lived in an apartment where we had lots of anonymous neighbors. One of the neighbors who watched this scene was concerned that they had just witnessed a child abduction and called the police. Granny hadn't made it a mile down the road when a police department patrol car pulled her over and demanded an explanation. This pales in comparison to what my Little Cupcake did to me several months later.

Back in those days I owned a white 1990 Cadillac Coupe de Ville with a blue carriage roof. I thought it was a sweet ride despite the fact that my wife told me I looked like a gay pimp in it. The car seat in the back didn't fit the car's personality. One day, my Little Jujube and I were running errands. We had made our first stop and were getting back into the car when my sweet 2- year-old went nuts. Despite baby's protest, I picked her up and stuffed her into her car seat and drove off. Being the dutiful parent, I had my Sweet Baby's car seat located in the backseat of the car, directly behind the driver's seat. As I pulled out of the parking lot and drove down the road, my Little Hell Bitch did non-stop leg kicks into the back of my seat. After 2 miles and more than 100 double leg kicks to the back of my seat, I tried to reach directly behind me in an effort to remove my Little Boo from her car seat to throw her out the window of my speeding gay cruiser.

Having your seat back jack-hammered by a pissed-off 2-year-old is not a pleasant experience. But I quickly forgot my rage when I was hit in the side of the head with a hard object. I turned around and was immediately rewarded with a blow to my face with the backseat window's trim panel. Little Love Muffin had torn the interior of the car apart and was using it to hurt her daddy. I played tug-of-war with baby's new bludgeon toy for a good half mile. After being kicked, hit, scratched and bitten, I determined there was no way I could get my hands on the fruit of my loins, so I pulled into a grocery store parking lot. I stopped the car, threw it into park, twisted around in my seat and made short work of removing my angry passenger. I put her out my open window and gingerly set her down on the ground. Her scream produced a 5-degree rise in temperature. I let her finish her tirade then told her, "You can't do this to me anymore. I'm your father and love you but I'm sick and tired of this evil shit. Sorry honey, but you have to live in this parking lot for the rest of your life." She took a step backward, straightened her arms by her sides, clenched her fists and inhaled all the air in the world as her eyes rolled back in her head. I was filled with a combination of pure dread and absolute fright as I drove off.

My Sugar Princess actually levitated several feet off the ground as I drove in large circles around her. I wanted to drive as far away as I could but my daughter was producing an energy force that had pulled my car into her gravitational field. Shopping carts, homeless people and other parking lot litter was blown into the air and formed a tornado of hapless debris around my Evil Little Mistress of the Universe. In the middle of her emotional super nova, Miss Cupcake pooped her Huggy, causing entire trailer parks of mullet-wearing white trash spontaneous bloody noses. This is when I gave up all hope and became completely liberated. It wasn't that I quit caring. In that moment I finally realized there wasn't a damned thing I could do about it. The only thing I could change was me; everything else was negotiable but ultimately out of my hands. It continues to be the strongest moment of clarity I've ever experienced.

In that instant, all the violence stopped. I was sitting in a parked American luxury car starring at the world's most beautiful 2-year-old. I asked her if she wanted ice cream. She smiled and nodded her head. I opened the door and leaned forward. She climbed into the back of the car, got in her car seat and buckled herself up. Nothing will make a 2-year-old more grateful or submissive than a clean diaper and an ice cream sundae.

These are the images that possessed my thoughts as the group of enlightened California fire-service professionals harped at me about requiring the members of the department to willingly join in on the forced march toward customer-service nirvana. I came out of my trance a stronger person. When the group finally took a collective breath, everyone had an expectant stare directed toward me. I looked over at my father. His look said, "Things were going to well. You just had to grab the wheel and run us over the cliff. Life was so much easier when I could do all of this with a slide projector." I made crazy eyes at Big Daddy before turning to the stirred-up crowd of earth people and confronted them with the real world. One occupied by B-shifters with large elbow and knee joints and hair-covered backs whose idea of high-quality service delivery had more to do with restraining themselves from physically assaulting the customers.

"I agree with everything we've been talking about for the last few hours. The leadership problem officers face has nothing to do with the highly motivated workers we've been talking about. But how do you train your officers to deal with the pricks in your organization? Those select few members who can't get along and refuse to go away. The idiots who leave a wake of pissed-off people. The vast majority of our citizens complaints come from rude and unprofessional behavior."

The Nordic goddess officer said, "As a boss, you cannot tolerate this type of behavior from the workforce. It must be stopped immediately."

"I agree. The members shouldn't be rude. How do you make them stop?"

"They must be disciplined until they understand the behavior will not be tolerated."

"That's swell. How do you do that?"

"When one of our members has a complaint filed against them, it must be investigated. If the complaint has merit, the member must be punished."

"When your department receives a citizen complaint, can you guess which one of your members generated that complaint?"

"It doesn't come as a shock. Most of our conduct complaints are generated by the same few people."

"Is it fair to say that if the same members continue to get complaints that your resolution process isn't working?"

A guy sitting in the back shouted, "How do you handle the complaints you get on your companies?"

I shared the following story: "I had a Captain working for me who was a world-class prick. This guy was angry at the world. Over the course of five or six months, he generated three citizen complaints and a complaint from a member of the department. The first two complaints I followed up on were filed by customers best described as clinically insane. One of them was a conspiracy theorist and the other one was a career criminal. Both complaints were so far out of whack there really wasn't much you could do with them. The third complaint was valid and so was the internal grievance. This gave me the opportunity to have a chat with Mr. Meanie. We sat down and argued for the first part of our get-together. He accurately pointed out that he was technically competent in the nuts and bolts part of his job. He just no longer had the patience to deal with all the "idiots and scumbags" who called 911. I agreed with him on both counts: He was competent and had no patience. I reminded him he was accumulating almost a complaint a month. Each one of these complaints pissed off half a dozen different people, ranging from the customer all the way to my boss, and I had to spend an average of three hours on each one of his human misadventures. I told him the complaints had to stop. He looked at me and with all sincerity said, "I've been here more than 20 years. What's the worst thing that can happen to me if I get another complaint? I'll still show up to work and get paid."

This is where I got to say, "You're absolutely right. I am not so naive to think I can fix all the things that are wrong with you. I also know that when I get another sustained complaint on you or your company I will spend an extra 30 minutes to write you up. After the second or third time this happens, someone will have to suspend you. I figure with an eight-hour investment on my part, I may be able to get you demoted. Life would be a lot simpler for all of us if you just stopped getting complaints. Every time you open your mouth, you piss someone off. If you stopped talking to the customers, you would probably stop getting complaints."

I was shocked when he told me the not talking idea had a lot of merit and he would give it a try. I left our meeting slightly more confused than I had started, but at least I got to put the cards on the table. I felt I had fulfilled my leadership responsibility to the organization. A week later I received another citizen complaint directed toward Captain Grumpy's crew. I called the complainant to get the details. She told me it was her intention to file a complaint against the police department for spraying her husband with pepper spray. Over the next hour, I found out the lady had dual citizenship (the United States and Germany). Her mother had been a secretary in the diplomatic core and her father was an officer in the Air force. She hated her across-the-street neighbors and she liked to play bridge. Her favorite singer was Tony Bennett. I asked her if she was happy with the service the fire department provided. She told me, "The firemen were very nice. I'm sorry they sent you my police complaint on accident. Please let them know they were very helpful in our time of need."

I exonerated the complaint, once again fulfilling my responsibilities to all involved parties. During the next four years that I was Captain Asshole's boss, he never got another complaint and as far as I know he never uttered a single word to a customer.

The guy in the back was having none of it. "You didn't fix the problem! All you did was make it worse by forbidding him to ever speak again."

"I'm going to have to disagree with you," I said. "My problem with him centered on citizen complaints. After our meeting, I never got another citizen complaint on him again. I fixed my problem."

"You didn't fix the organizational problem," he continued. "I was an officer in the military, and if I had this problem I would have…"

I stopped the guy and reminded him that the military can imprison and execute its troublemakers. I was just a lowly battalion chief trying to get through the day.

The guy in the back was now standing and speaking loudly. The well-behaved audience had dissolved into a dozen or so independent talk groups trying to shout over one another. I had turned them into B-shifters. Finally the fire chief stood up and declared break time. The time-out allowed the group to find its center, restoring the harmony and free love that I interrupted. I spent the rest of the afternoon nodding like one of the soldiers of Islam that stand behind Louis Farrakhan when he makes his speeches. All I needed were some dark shades and a bow tie.

Sometimes we overestimate our personal power. Other times we confuse any use of our power as good leadership. People follow other people because they want to. There are a million different ways to lead because there are a million different motivations that cause people to fall into line. Money, power, being included, continued employment, good dental benefits and ice cream are just a few of the things we will trade for our compliance. Captain Asshole traded letting the public know exactly how he felt for his own future peace and status quo. My 2-year-old daughter taught me that you shouldn't get too caught up in the things you can't fix. Sometimes it's all you can do to get out of the way and not make things worse. If you feel like it's your inalienable right to control the world and people around you, you'd better buy a helmet and look forward to your daily ass-kicking.

Chapter 26
Hate & the Preacher

One of the topics considered most taboo in the fire station is religion (unless you happen to be the chaplain). Religion is a touchy theme because it is a very personal subject, rife with eternal implications. This can lead to emotionally explosive situations in fire stations (not to mention a 2,000-yearold holy war in the Middle East).

There is a reason our founding fathers separated church and state. Managing the routine duties of a fire company is challenging enough without having to deal with each member's faith-based beliefs. One of the drawbacks to the religious debate is it ultimately segregates our very essence (our souls, if you will) into two groups: the chosen and the damned. Pick the right god and you're a winner: Every hand is blackjack, a herd of 72 black-eyed virgins awaits you and you get a fat signing bonus to lead the A-shift softball team. Pick the wrong god, and you're a loser: spending eternity drinking dead baby milkshakes, enduring an endless tax audit and receiving a fat signing bonus to lead the A-shift softball team. The last thing we want in fire stations is to divide the 24-hour workforce into winners and losers. This is a tough enough thing to manage at shift change when the A-shifters have to mingle with the other two shifts.

I have a colleague that had just secured a choice spot at a single engine company house. He was looking forward to spending the next four or five years in this spot while preparing for a future captain's test. I ran into him a few months later and asked him how it was going. I was mildly surprised when he told me, "I only worked there for a couple of weeks. I'm out roving again." This threw me off balance. "You've been roving for the last three years, why would you give up a good spot after just a couple shifts?" He hemmed and hawed with an answer. When I suggested his departure was the result of a torrid love triangle, he finally came clean. "I left because the three other guys were Jesus freaks."

This was a real let down for me. My head was bursting with visions of a heavily breasted female getting lathered up watching firefighters locking horns and spraying one another with urine. "That's it? You left because they believe in the Christian Son of God?" My friend was now letting those tortuous few weeks seethe back into his demeanor. "You got no idea how horrible it was working with those three. I didn't leave because we said grace before every meal. I bit my tongue every time I walked by a television set with that fascist Pat Robertson reminding me that I'm personally responsible for nailing Jesus to the cross. I gave my spot up when they began preaching on calls. Do you know how uncomfortable it is to be taking a patient's blood pressure and have your captain ask him 'Sir, have you accepted Jesus Christ as your personal lord and savior?' After we got back to the station I tried to talk to my captain about his inappropriate behavior. You ever try to talk reason with a religious zealot? I was the odd man out. Leaving that place was the smartest professional decision I've ever made."

After my friend left, I sat and pondered a solution to

his problem. I reasoned that his situation boiled down to group dynamics. It was Darwinian, when 75 percent of the group shares a common trait they will overwhelm the minority. My buddy was right—he was the odd man out. Before some of you start getting all heated over the need for organizational intervention and brining down the boot of the emperor on these evangelical emergency responders, let me share an episode that swung the other way. In the mean time, those of you thinking the God squad is off base are correct. If these guys want to save souls they can buy bicycles, black pants, white shirts and skinny ties and do missionary work on their days off.

A long time ago I roved into our department's southern most station. It was nestled in a rural area full of Japanese-owned flower farms and roadside stands that sold date and pineapple milkshakes. The first-due area was inhabited by a diverse group of working-class people who seemed to live at a slower pace. I can still vividly recall the large vacant field behind the station filling up on the weekends. Hispanic oompa-loompa music would fill the air along side the smell of cooking meat. By mid-afternoon, hundreds of people would be wagering on the dozen or so cockfights taking place. At the end of the day, the revelers would clean up their mess along with the dead fighting birds, pack the family back into the Chevy truck and head home, ready for another week of day labor.

One of the firefighters assigned to this station was an evangelical preacher. He was a brawny, raw-boned man with big hands and a greased crew cut. His side job was ministering to a flock that assembled in a tent at the base of South Mountain. His major problem was that he had trouble turning it off at the fire station. One day the two of us were riding on the tailboard of the truck when out of nowhere he told me, "Bruno, I'm the luckiest man in the world. Do you know why?"

"No Ray, why are you the luckiest man in the world?"

"Because when I die I will live in the kingdom of heaven with Jesus Christ."

"Ray that would make you the luckiest man in the world when you die. I am the luckiest living man in the world. Do you know why?"

"No Bruno, why?"

"Because when I get off work in the morning I get to play with a big, nasty red head's stinky winky." This effectively ended all chitchat between the two of us for the remainder of the shift.

Reverend Ray used his size to bully his religion on the unsuspecting members of the fire department. Firefighters tend to be good natured and patient when meeting a new member. Most guys would put up with a salvation-filled shift and ask not be sent back to Ray's station. Most guys will put up with it, but not all guys will. Ray had the unfortunate luck one morning of coming across a man who was more comfortable playing skip rope with the devil than sitting next to Jesus on a cloud pillow in heaven. I had the great joy of getting to watch the freak show dance between Reverend Ray and Big Daddy Hate.

Big Daddy Hate's work experience included automobile repossession and being an associate member of a motorcycle "club." He always had a mischievous glint in his eye. People who didn't know him would describe the twinkle as the same a 12- year-old gets after receiving a BB gun from Santa. Those who knew Big Daddy on a deeper level recognized it as the look of anticipation a rogue alligator gets just before it taking down an unsuspecting jogger.

I showed up to the little southern station and found the captain in the day room. He was alternating between his morning coffee and strumming his well-worn banjo. The man fit the first-due area like a glove. We were in the midst of a shared morning greeting when a shriek exploded from the back of the station. The old captain looked at me and said, "Looks like Ray just met that Hate guy. Never worked with Hate but I hear he's a fine engineer." He took a sip of coffee and went back to strumming hillbilly music. I ventured into the back of the station because I'm curious as a cat. The first thing I noticed when I swung the door open to the station dorm was the Reverend huddled in a corner. His face wore a mask of abject fear. His mouth was moving, but it wasn't making any noise. He looked like a fish dying on hot concrete. Big Daddy Hate was standing over him, giggling like a little girl. Before either of them noticed my interruption, Big Daddy purred, "Come on Preacher Man, touch my little Jesus. He wants to baptize you."

When the Rev took notice of me he blubbered, "Make him stop. Please make him stop." Big Hate turned and smiled at me with a hand full of his baby maker and said, "Come on Nick, there's enough of him for the two of us."

Ray took advantage of this brief distraction and bolted from the room. Big Daddy stuffed himself back into his pants and went to check out the rig. I wandered back into the dayroom and found Ray chattering away at the old hillbilly captain. Ray wanted something done about the jailhouse sexual assault that Big Daddy Hate almost perpetrated on him. The old captain stopped playing his banjo long enough to say, "Ray, who in the hell told you it was going to be easy saving people's souls?" The old captain then went back to his pickin' and the rest of us fell into our normal routine like the event never happened. Ray quit preaching and Big Daddy Hate remained fully clothed. The problem just went away.

This event happened during an earlier time. It was an era when the local population could assemble on vacant property to have a barbecue and wager on poultry fighting to their death. In the fire station workplace, if someone did something you considered obnoxious you could return the favor in kind. Back then we all just took care of our own business. Big Daddy Hate had grown weary with all of Ray's daily affirmations and was putting an end to them. We all knew it was just theater (OK, maybe Ray didn't). There were no phone calls made, forms filled out, personnel chiefs notified or lawyers brought in. We didn't inflict a department-wide sensitivity training session on all of our members. Right, wrong or otherwise, it's just the way we handled things back in those days. In the end maybe its every bit as wrong to intimidate co-workers with your genitals as it is to preach to them. A lot of stuff has changed in our professional lives during the last two and half decades but one thing remains constant: Ministering is best left to the chaplin.

Chapter 27
IRS-Fueled Cults

Since this is tax month, I would like to devote this article to those intrepid firehouse accountants. I think most of us have at least one of these crafty individuals working in our departments. I'm not talking about those members who actually went to a university and earned a degree in accounting—you can find these qualified accounts on the Internet, or simply get referred to a top-flight accounting professional by a friend or associate. I'm talking about financial bulldogs. You know who they are. Guys who have spent hours figuring out loop holes and researching endless hours on the internet, finding proven ways to hoodwink the government into giving you back all of your hard-earned money. Let me begin with an early tax visionary.

When I was a young firefighter (earning around $20,000 a year), I spent a few shifts with a captain who was a tax-preparation pioneer. This individual was a very nice gentleman, and he was a pleasure to work for. During our first shift together, the crew was eating lunch when the other firefighter (also a young fill-in guy) began lamenting about having to get his taxes done. It was the typically rant: "The damned government has already fleeced me enough. For the love of God, how much more do they want! Will the tyranny never end?" After the young firefighter finished, the captain looked over his half finished cheeseburger and asked, "Would you like me to help you with your taxes?" Looking back, I'm sure it sounded a lot like the time the snake asked Eve if she wouldn't like an apple.

Not having an accountant of my own, I asked the captain over lunch dishes how long he'd been an accountant. He told me, "I started writing computer programs last year and got bored with it. I picked up a couple of books on tax preparation and joined a tax revolution club. I've been doing taxes for a few weeks now. It's not as tough as it looks." It is important that you younger viewers keep in mind that at the time this event took place no one knew who Bill Gates was and the personal PC really wasn't a mainstream thing yet. Half of the personal vehicles in the fire-station parking lot had cassette decks. The other half were older and had 8-tracks. The captain did have distinguished graying at his temples and a dazzling smile. It was easy to place one's financial faith in his hands.

After the noontime dishwashing was completed, the captain got his brief case from his office and sat down at the freshly wiped kitchen table. For some reason I can vividly recall the leather briefcase sticking to the dishwater swirl patterns left by the sponge. With the other young firefighter taking his place at the table, the captain asked me, "Bruno, grab some paper towels and sit down. I can help you with your taxes, too." I grabbed a handful of paper towels, cleaned up the traces of water and sat. The captain thumbed the secret code into the latches of his brief case and popped open the lid. The first thing I noticed among all the paper, forms and tax books was a Colt 45 semi auto with four extra clips. I have to be honest—I don't know very many software engineers or accountants, but the ones I do know don't pack heat. The captain saw me looking, smiled and said, "I do business all over town. A man can't be too prepared."

The captain retrieved the proper form from his briefcase, looked at his new client and asked, "Okay, how much you want back?" The young firefighter looked at his captain-accountant with confusion before replying, "Excuse me sir, but don't you fill out the top of the form first and work your way down? You know, stuff like my name, address, social security number…?" Before the youngster could finish with his question, the captain cut him off with a shake of the head and wave of the hand, "Son, this is a radical new approach to getting back your money. I do not take this social responsibility lightly. This tax season alone I have earned hundreds of my clients over half a million dollars in tax returns. What I do is perfectly legal, and until the IRS passes new tax laws to stop me, I will continue to rape them in the same fashion they have been raping our society for the past 200 years. I guarantee you that my methods are bulletproof." He and his newest client spent the next 15 minutes working backward from an $8,000 refund. The ensuing audit shouldn't have come as a surprise since the young firefighter had only paid half that amount in total tax.

I don't know if it was just a local phenomenon in the little microcosm that was my department, but in the early 1980s we had more than our share of tax revolutionaries. Many of these individuals became ordained ministers in fringe churches. According to the ordained believers, one of the benefits of becoming an official man of God was that your home became God's home. All you needed to do was buy some folding chairs and have a weekly party where you and your friends (mostly other new ministers) could discuss spiritual matters and different strategies to fleece the government out of their forced tax contributions. This practice became widespread enough that the IRS wrote special tax bulletins to their auditors and agents warning them to be on the lookout for the recent influx of the newly formed churches that were run out of firefighter's homes. The early '80s were truly a far-out time, but the IRS has never been known for their patience or having a sense of humor. The journey to tax salvation ended badly for these tax-avoidance pioneers. All of these ministers of tax avoidance were done in by legal battles, threats of incarceration and ultimately an IRS garnishment that would have made the most rabid ex-wife envious. It came as no surprise when these broken men sought refuge in another fringe hobby to help rebuild their self-esteem. The fake preacher men sought comfort in a cult called "Life Spring."

One of the members of this society of insight and enlightenment was an African American officer who was fond of wearing capes (I'm not making this up) and carrying a walking stick. He looked like he'd just fallen off a label of Johnny Walker whiskey. He would routinely rendezvous at shift change with several members of the station I was assigned to. These other members were mostly white, poorly dressed tax evaders (actually they were failed tax evaders). They would gather in a corner and speak in hushed whispers before leaving for that day's meeting of power. One day, the Life Spring brother was coming through one of the bay doors to meet up with his fellow cult members when I ran into him. I was admiring his faux fur cape and chrome walking stick. I said, "Why Demetrius, aren't your looking extremely dapper today. You look like the steering wheel on a marching band. Why do you hang out with those lame tax protesters upstairs? What do you guys do at these cult rallies?"

He smiled and said, "That's easy Bruno Jr. There is an incredible amount of confused, fine-looking pussy at these pep rallies. They all cravin' the touch of a strong man."

Living on the edge isn't just fun and games. You've always got to stay on your toes when you stray too far from the accepted practices of the herd. Between the government, God and the competition, there is always somebody gunning for you. One final suggestion—never take advice (of any kind) from a man who wears fur capes.

Chapter 28
Hillbilly-Gypsy Neighbors

A major component of living a rich and fulfilling life involves dreaming big dreams. Some might say our capacity to dream makes us human, maintaining that life wouldn't be worth living if we couldn't dream. Dreaming sets our purpose in life and gives us something to strive for. On the other hand, people who have big dreams sometimes cut off their ear and send it to a former lover. (Life is all about balance.) If big dreams provide the highest of highs, then on the flip side they can lead to a darkness so absolute, it makes you forever insane.

I have been a family man for more than 25 years. This doesn't count my early family experience as a toddler—created and raised by a married heterosexual man and woman, ultimately becoming a sibling to a brother and a sister. I am also excluding the period in my life when I was freshly married up until the time my wife had our first child. During this "beginning" period of my life, I was primarily devoted to my own amusement and self-gratification. During these "me" years, I didn't do a whole lot of big dreaming because as a standard course of life I did pretty much what felt best within the wide margins of modern society. I was young; work was fun, and it financed all the consumption, touching, tasting and intoxication I could fit into a day. The future was something old people and the well-pressed spoke about. All this changed for me when my betrothed dropped our first foal. My adventure into the world of big dreams began when the overriding focus of my days became feeding, clothing, housing and maintaining a family unit.

After being blessed with our third child, we determined our 1,600-square-foot house would be quickly outgrown. We had big dreams of building a new home on a small mountainside lot that bordered a park. Our big dreams prompted us to purchase the lot, sell our small house, move into an apartment and begin construction on our new home immediately. Our dream forecast had us spending no more than a year in the cozy confines of our new 1,100-square-foot apartment. We reasoned the pains of living in a smaller abode would easily be anesthetized by the joy of building our own little hillside Valhalla. If I have any advice to offer, it is this: Never attempt to build a custom home when the neighborhood association is suing the developer who owns every other available lot and has successfully ceased construction on a half dozen houses. It took you less than a 30 seconds to read this advice; it took me six months and thousands of dollars to figure this out.

Dreams are nice, but reality always plays the next card. What began as a dream to build a new house turned into 18 months of apartment hell. We lived on the second floor of a three-story building. It was in one of the cities higher-rent districts, a few blocks from the patch of the dirt that was supposed to become our home. During those cramped months, I never recall hearing the people who lived above me or next door to me. They could have been making more than their share of noise but you'd never know it. They would have to been testing jet engines or castrating rabid baboons to be heard over the people who lived directly below me. During the infancy of our relocation, my upstairs neighbor knocked on my door one night to ask us if we could keep the noise down. Before he was able to finish his request, he was interrupted by shrieking and slamming doors on the ground floor. He apologized for the interruption and headed downstairs to plead for silence from the downstairs Satan worshipers. I wished him luck. This was the first and last time I ever saw my upstairs neighbor.

Several months before my family moved out, I struck a blow for every frustrated tenant in our apartment building. My wife and daughters had gone to a chick flick, leaving me alone with our 200-pound dog for an evening of checkers and the passing of the foulest-smelling gas any mammal has ever emitted. The dog and I had grown numb to the constant hum of arguing from down below. The evening's normal volume was breached when one of the occupants of the lower unit shrieked, "I hate every one of you! I'm going to kill myself!" This opening had the same effect as ringing the bell for a cage match between a dozen professional wrestlers. Within seconds, I was trapped in an unedited version of the "Jerry Springer Show." The downstairs matriarch shouted, "Quit your whining, you little titty baby. Why don't you try getting up off your worthless ass and getting a job? I pay the rent here…"

The dog looked at me with weary eyes and pushed out a load of gas that contained enough sulfur to kill a water buffalo. I sat back on the sofa, looked up at the ceiling and took shallow breaths through my mouth and wondered out loud, "What about my dreams?" This caused me to reflect on my life. I had long since given up on my dream of moving a 17-year-old Swedish girl in as an au pair. I was trying to focus on my internal sadness over the loss of a young blonde with champagne-glass shaped breasts giving me an evening bath in between doing loads of laundry and reading my children to sleep. I couldn't wallow in the full depths of my misery because my God damn downstairs neighbors were shrieking at the top of their hillbilly lungs.

The dog was barely interested as I climbed up on the coffee table in the small space that served as my living room, dining room and home office. The emotional riot downstairs was in full swing as I crouched down into a squat. Mama Sea Hag continued shrieking at the top of her lungs. It sounded like bad men were ripping her eyes from their sockets. Deciding that I would no longer suffer my humiliation in silence, I leapt off the coffee table with the strength of a samurai. I felt my head bump the ceiling before gravity began pulling me back toward the core of the earth. Prior to touchdown I drove my legs as hard as I could into the floor. The force of my landing shook the entire building and caused tremors as far away as Riverside, Calif. This was immediately followed by a silence that I hadn't heard since I moved out of my too-small house. Several minutes later, there was a knock at my door. I opened it to find one of the 23 occupants from the lower unit.

He was in his mid-20s, stood at 66 inches and weighed

approximately 105 pounds. In his hands were several of the plastic sleeves that go over the sprinkler heads that protected his unit from fire.

I looked at him and said, "You knocked?" He blurted, "I don't know what you're doing up here, but you knocked these off our ceiling. We would appreciate it if you could keep it down."

"You have got to be shitting me," I replied. "You people make more noise than a Civil War field hospital. You got some balls coming up here and rousting me and Dog on checkers night."

He peeked around me and looked at the large, stinking dog sitting behind me. When his gaze found me again, he said (in a very put-out but slightly afraid tone), "We've just had a death in the family and don't need you jumping around and upsetting us more than we already are." I can't be certain what caused me to say what I said next; maybe the realization that I would never have a 17-year-old au pair finally hit me. "If you interrupt Mr. Stinky and me for the remainder of the evening the two of us will come downstairs, clog your toilets, break your TV, and Dog will breed with every one of you high-strung mountain people over the age of 18. When we're done I guarantee you will never get your cleaning deposit back, and your entire family will walk with a limp for the rest of your meaningless lives."

Before the insipid little noisemaker could respond, I slammed the door hard enough to make him sterile. I went back into my humble little abode and fed Dog four pounds of leftover birthday cake and a large bowl of buttered corn. The next day, Dog left the vilest, most repulsive and artistically textured dog gift in front of the downstairs neighbor's front door.

This brings me to the point of my little essay—relationships with one's neighbors. Over the past five years, a lot of time, energy and money have been thrown at developing the National Response Plan. This plan has gotten lost in all the big federal NIMS broo-ha-ha. Given the choice between reading the National Response Plan and extracting an impacted tooth with a rusty pair of vise-grips, most sane people would pick reading the federally generated plan. The average firefighter wouldn't get halfway through this reading assignment before asking if it wasn't too late to opt for the pliers.

Most of the National Response Plan (and NIMS) focuses on how each federal agency should work with all of the other federal agencies. In the event of an emergency, the Department of Housing reports to the Department of Agriculture, who reports to the Department of Interior. The Department of the Interior is tall dog until the FBI shows up and sets up a command post with a roof over it (a JOC in the Box). If the FBI sets up an open-air command post, then FEMA will remain in command of the event. If FEMA remains in charge of the event, we will have to fill out enough forms, requisitions and other paperwork to fill half a dozen filing cabinets in order to obtain any permissions, assistance or help of any kind. These forms will languish in some federal purgatory while FEMA overnights checks to felons doing hard time and finances sex-change surgeries for individuals who are

238

in no way affiliated with any emergency-response agency. The inhumanity of it all.

I have a simple solution. When we respond to an incident, we send a standard amount of resources based on the call type. When the first-arriving unit shows up to the scene, they perform a size-up of the incident. This size-up includes determining the level of required resources. If the incident commander feels more units are required, he/she requests more units by striking additional alarms.

All of our departments have a finite amount of additional resources. When the incident escalates beyond the resource capability of our department, we typically will request the assistance of neighboring fire departments. This is where life gets a little fuzzy for much of the fire service.

The most effective aid agreement between fire departments is "automatic aid." "Automatic" means when the IC is sitting in front of a working incident and keys his/her radio mic to request more help, it automatically happens regardless of where the resource comes from. The IC is sent the closest, most appropriate resource. If the incident is occurring on the border of two cities, the initial response will typically be a mix of resources from both departments. Automatic also means that in the event the first-arriving unit is from an outside department, they will take standard initial action and assume command of the event. This same logic also holds true if the first-arriving command officer is from a different city. In this case, he/she will go ahead and take command of the event just like they would if the incident were taking place in their own city. In a few spots around America, it is not unusual to have half a dozen or more different fire departments operating together like one big, happy response family at the scene of a large local incident.

This is not the case for much of the American Fire Service. A lot of places use the mutual-aid agreement approach to helping their neighbors. It is not unusual to hear firefighters say, "We don't mind responding with the guys from Happy Town but we will never respond into Zazaville. Those guys in Zazaville are all a bunch of ass clowns and their Mayor is a vegan. Last year he picketed our barbecue fundraiser. He ever shows his face around here, we are going to force-feed him an entire ham." When you listen to the details of some of these dysfunctional relationships, it is easy to believe that it wouldn't matter if it was raining giant burning frogs and the tactical channel was filling up with maydays—they would not call Zazaville for help.

There are just as many mutual-aid opportunities where firefighters from bordering departments want to help one another but are stopped by bureaucracy. In most of these places, the dysfunctional relationship exists at the fire chief level or higher. When the powers that be attempt to explain the convoluted process required to send help (the mayor from one town has to call the mayor from the other town and then they have a conference call with the governor and the chairman of tow-truck drivers), it ends up sounding like they're reading from the National Response Plan. In many cases, this process takes so long it quits being a request for outside help and becomes a heads up that the conflagration has reached the town border and the problem is about to be shared equally by both communities.

What would happen if it became a federal mandate that every fire department had to run automatic aid with every fire department that shared a geographic border? We would all retain our own unique identities: silly uniforms, apparatus color, grooming standards, kitty fees and rank structure. Things would begin to change in our operational world. We would soon find out that we had a lot more resources available to us in a timely manner (like when the building was actually still on fire). We may even begin to train together, ensure our radio systems are compatible and use one another to meet our mutual needs. A department needs a new tanker and a brush truck. They can't afford both. If their neighboring department buys a brush truck, they can buy a tanker and both can share them. Imagine if the 30 departments that coexist in a 200-square-mile area all got together and purchased SCBAs as a single group. It is a pretty safe bet they would all realize big cost savings and more attentive customer service. Multiply that by more departments and all the other stuff we purchase every year. The fire service would end up with its own QVC.

When we begin to operate with our neighbors as a regular part of delivering our day-to-day service, it establishes a framework that connects all of the departments spread over many communities. Responding together as a regular part of our local service delivery creates the strongest possible foundation. In the event the big one hits town, we will be in a much better position to respond because we are hooked up, well-practiced and it all happens automatically. When I'm managing a hazard-zone operation, I couldn't give a shit where the fire trucks and firefighters come from. All I'm really concerned about is that they're equipped, trained, and ready to go to work and I can communicate with them. The 20+ departments in my corner of the world have spent the better part of the last 30 years making our individual departments connect "automatically." If all of the local fire departments in our country did this, would we be in a better position to engage the federal authorities regarding what we want and need? I guess a better question is, If all America's neighboring fire departments united in a nationwide emergency-response system, wouldn't that make us the only real federal emergency response system? Wouldn't that be dreamy?

Chapter 29
Barber-Based EMS

A couple of months ago I was going through some boxes in my garage and ran across an old uniform jacket. It was the jacket I wore when I was the captain of a ladder company that was named after a gay pirate ship. It filled me with strong memories. As I lovingly caressed a piece of my past I noticed the pockets contained little treasures from that bygone era. The right front pocket was stuffed with blue latex rubber EMS gloves. I vividly recall handing out gloves just like these to the prostitutes as public service. Most of them would giggle when told they were special, cinco use condoms. The other front pocket contained a book of taxi vouchers. Many of the people that accounted for our company's EMS related encounters preferred a free taxi ride over the use of an ambulance.

I had learned a valuable (not to mention cost saving) lesson to always fill out the destination portion of the taxi voucher prior to handing it over to the customer after a voucher I issued was used to take a $300 cab ride to Tucson. I should have known better than to give an open transportation ticket to a half naked man that we found running around in the middle of traffic, spitting on motorist's windshields. After the taxi bill was brought to the attention of the sinister forces that ran our budget department I was drug before my chief for an interrogation. In a misguided effort to lighten the mood I said, "Chief, were not looking at the positive aspects of this service delivery encounter.

Any true Phoenician knows it is priceless whenever we can ship our undesirables to Tucson. Who knows, maybe we did them a favor. The Phoenix Fire Department may have sent Tucson their next Mayor." The chief scowled as he informed me that his daughter was attending the University of Arizona and resided in the city of Tucson, like that was somehow my fault.

The found object that rocketed me back to the good old days had been stored away on the inside pocket (or the secret agent slot) of my old uniform jacket. It was an innocent little green box, containing 10 ammonia inhalants. It still had its cellophane wrapper intact. Before we became a kinder and gentler customer-centered, service-delivery organization the ammonia inhalant was better known as the "oxygen pill." It was a very effective tool for dealing with patients who felt the need to fake medical symptoms ranging from unconsciousness to death.

These self-induced fainting spells screwed up the flow of the call. We would show up and typically move through a standard set of phases for these lower tier EMS calls. The process began with the introduction ("fire department") before moving onto the patient interview. We then commenced with the taking of vitals and the filling out of the form prior to moving into the treatment phase. When the patient threw a fake swoon in the middle of this process we would become stuck, often times between phases, which in severe cases caused us to begin the whole thing all over again. I don't know about the EMTs and Paramedics in your department, but back in those days we were overachievers and our time was too precious to be wasted on people going into fake shock.

We also reasoned that you really couldn't go wrong with the administering of the ammonia inhalant. If the patient were faking it, the magic pill would snap them right out of it. On the other hand, if the patient had truly died they were already dead and it has been proven by government scientists that ammonia inhalants have no lasting side effects on the dead. We couldn't loose. When Fire Chief Alan "Customer Service" Brunacini heard about a radical new delivery system for the administering of ammonia therapy he eradicated the oxygen pill in our department forever.

Your average ammonia-inhalant candidate didn't have any medical problem. Most of our ammonia therapy was prescribed for patients who just had some type of negative emotional encounter with a loved one. I've heard this affliction described as "emotional hysteria," "chicken breeders," "female fatigue" and the ever popular "peckerwood just fell out." It seems that each socioeconomic and ethnic group calls it something different. It is my advice not to use these street terms for documentation purposes on an EMS form. The medical forms we fill out are legal documents and are typically reviewed by medical professionals. The things you and your crew find amusing at three AM will get a different reaction when read by the regulators, patient and family members, bureaucrats along with the legal world that routinely reads our reports. You should make it a point to avoid writing anything down that causes you a forced audience at the Tower of Doom (in a dress uniform no less) to discuss your future employment status.

Ammonia therapy was a highly effective treatment intervention. It worked virtually all of the time. One of the problems with ammonia, as with life in general, is there is no absolute. Occasionally we would encounter a patient who could outperform the ammonia. These artists had the respiratory capacity of old school, Japanese pearl divers. They could literally suspend their breathing for the length of time it takes to hard boil an egg. I heard a first hand account from an A shift crew that had diagnosed one of these faker/breathing savants as actually having gone into respiratory arrest. The patient was exposed as a fraud when the medic began the invasive intubation procedure. The prevailing technique for checking the patient's mental status was to pop a couple of inhalants and cup them over their nose while covering the patient's mouth. I learned this technique from an ER doctor. It should come as no surprise that a firefighter took it to the next level.

One piece of EMS gear that we used quite frequently was the bag-valve mask. It has been over a decade since I've worked on a piece of apparatus that carries EMS gear. I'm told that we still carry this lifesaving piece of equipment on all of our rigs. The ensuing 10-year evolution in bag-valve mask technology has made today's model single use and disposable. I have fond firefighter memories of cleaning the bag valve mask in the bathroom of Fire Station 21. One would rinse off the outside of the instrument prior to unscrewing the end cap (where the O_2 connection was located). After opening the football-shaped billows, the vomit, bone, pieces of lip, and teeth would be rinsed into the bathroom sink. After scrubbing the contaminated human destruction from the tool it would be soaked in a bucket of commercial disinfectant for 30 minutes.

While the device soaked, the human fodder that was too big to fit down the bathroom sink drain got thrown in the trash. This procedure was performed bare handed as the only rubber gloves we possessed were hermetically sealed in the OB kit. It was against department policy to touch any part of a pregnant woman's vagina without the protection of a sterile rubber glove. I'm sure many of you younger readers are thinking to yourself, "Not more of this good-old-days claptrap." Not to fear kiddies—listening to the ramblings of the aged and insane just goes with the fraternity you joined. Besides, on a far-away day you, too, will be bombarded by the artillery of the past when you accidentally stumble across a talisman from your long-forgotten time. Back to the ammonia-inhalant story.

One of the problems with the ER-doctor-taught ammonia-inhalant method was the occurrence of being bitten. The fake fainters could not hold their breath long enough to avoid inhaling ammonia. In retrospect it doesn't make much sense to put doses of pure poison in the EMS kit—even for a government agency. Once the toxic substance hit the "patient's" sensory organs they would commit any act to be rid of it. This includes biting any obstacle that stands between them and fresh air. Bite me once shame on you. Bite me twice shame on me. The solution to the biting problem was to modify the delivery system. The bag-valve mask delivered 100 percent pure oxygen to the dead and dying. Why couldn't we use the same medical tool to deliver highly concentrated ammonia to people with a questionable level of consciousness?

Was it wrong? Was it effective? If its creator had the opportunity to go back in time to un-invent it would he? The answer to these questions almost certainly must be yes but I reserve the right to refer back to an earlier passage regarding life containing no absolutes. The Ammonia-Ambu-Blaster was destined to become outlawed. It was one of those things that was never sanctioned, authorized, or institutionalized. As far as I know it was just another piece of equipment in the B shift toolbox. Like any fire company worth our salt, we would teach young firefighters fresh out of the training academy a set of skills that would make their professional life more exciting and fulfilling. One of our former apprentices heeded the lessons well and found himself having to explain his actions in front of a full-blown EMS tribunal.

The incident involved a female patient in her late teens. She just went through some type of family/boyfriend/ baby's daddy emotional drama and the fire department had been summoned because she was stricken with a severe case of emotional hysteria. Our former probie was several years into his career and a key member of the responding crew. The crew arrived to the scene and was in the process of delivering standard, high quality emergency services to the entire assembly of patient and family. It was one of those calls where none of the principles know what happened but that in no way interfered with each one of them having a conflicting story to tell. Less than 10 percent of what the patient said was comprehensible English. Most of her communications were excited chatter, grunts, and sobbing pig Latin. Several minutes into the event, the patient was frantically attempting to explain herself when she went limp and fainted into the over stuffed sofa. As the crew attempted to rouse her they were called away by a family member with a fresh set of false information. The patient sensed she was no longer the center of attention and made a miraculous, yet brief recovery from her near death experience. Several members of the crew returned to the hyper-alert patient. Within minutes the sobbing resumed followed by another total collapse of her central nervous system. This same pattern was repeated several times.

We are not sadists. The crew used medically accepted exam techniques to determine the patient's neurological status. When they performed the eyelash stroke the patient fluttered her eyelids. The hand drop maneuver resulted in the patient's hand taking a controlled path away from her face every time. The crew warned the patient that they would wake her up and she would talk to them. She only fell deeper into her fake coma. The next step was to pop a single ammonia inhalant and place it on her top lip. After a few seconds this caused the comatose patient to roll to one side and lapse into a fake seizure. The choreographed convulsions lasted a minute or two before the patient fell back into the clutches of another coma. Our young charge remembered the effectiveness of the Ammonia-Ambu-Blaster for curing individuals with fake comas. The crew looked on with the curiousness of young boys as the junior guy on the crew loaded and prepared the device. The crew quickly figured out where this was going and assumed patient hold-down positions at the shoulders and legs while the young firefighter administered the cure. One squeeze of the bag produced a coughing fit that was followed by the foulest cursing that several members of the crew had ever heard (they were both Baptists). At this point in the service delivery encounter the police had shown up. One of the officers obtained the patient's driver license and determined she was wanted on warrants. Our soon to be former patient had a coma relapse when the police officer informed her she was under arrest and began to read her her rights. The cop looked at the crew and said, "It appears she just went into arrest-o-genic shock. Is there anything you guys can do?" The young firefighter replied in a happy voice, "Sure, we can give her more gas from our magic bag." This had an immediate curative effect. The patient got to her feet and proclaimed, "All of you are a bunch of assholes" before marching into the darkness of two AM and the back of a waiting PD cruiser.

As fate would have it, that evening's patient was a distant relative of a member of our department. After hearing the details of the story the member was less than thrilled with the new treatment protocols B shift invented. After half a dozen well-placed phone calls a kangaroo court consisting of the chief of EMS, the fire department medical director, a couple of senior A-shift medics and an angry nurse was assembled. The doomed crew sweated through their dress uniforms as they waited in the hall. The first part of the meeting went pretty much as you'd imagine. It was an outrage, a tragedy, and the end of the free world as we know it. The crew pleaded their case. They never meant any harm to patient but a member of the family had told the crew that she watched the patient consume an unknown amount of brightly colored pills. The crew had the need to question the patient and she wouldn't cooperate. "What were we supposed to do?"

Half way into the meeting things took a funny turn when the medical director said, "You guys have a difficult job. You did everything perfect with patient care right up until the time you ventilated her with a high concentration of ammonia. On the other hand, we supplied the ammonia inhalants as part of the standard medical supply inventory."

The EMS chief wasn't buying it. He screamed, "They should have never done what they did, and you can't make excuses for them. What if they had given her a Drano IV?"

For all practical purposes the meeting was over when the medical director told the chief, "We don't stock the drug boxes with Drano."

This is one of those instances where everything ended up just right. The medical director met with the fire chief and they both agreed it was time to bid farewell to the oxygen pills. None of the ammonia administering crewmembers were flogged, flayed or written up. The next department EMS training block covered strategies and interventions for dealing with patients who slip in and out of fake seizures. A few short years after the ammonia incident our fire chief grabbed the organization by our neck and dragged us into the brave new frontier of fire department customer service. In the mean time I continue to be fascinated and hypnotized by some of the stuff I find in my garage.

Chapter 30
The Middle-Child Rank

The other day I was lunching with several of my rabble-rouser buddies. We were shooting the breeze about the current state of affairs in our corner of the universe when the discussion took a left turn down memory lane. The senior member of the group asked the innocent question, "During the course of your career, what's been your favorite position?"

"Doggy-style because I like to play with her hair."

"That's a load of shit. Everyone at this table knows you prefer doggy-style because your gay lover's lumberjack-beard irritates your sensitive neck skin when you engage in missionary position intimacy."

The philosophical one of us said, "You two are uncouth barbarians. Lately I prefer to be on the bottom, that way I can close my eyes and think of England."

This went on for several fun filled minutes before the author of the original question regained the floor, "I was asking about work, not your fantasy lives. What was your favorite rank? Where was the best place you've ever worked?"

The answers ranged from firefighter to shift commander. I don't know if it's possible to pick a best but I've never had a better time than when I was a battalion chief. During those heady days my partner and I had a response rig and a battalion. We didn't fill out forms, hold needless meetings, or wear dress uniforms. We ran calls and hung out with the companies in our battalion. At the end of our shift there was no administrative proof that we existed. We had all the authority but weren't responsible for anything that fell outside of managing hazard zones. My boss was a nice guy who I saw half a dozen times a month. It was a simple time.

As the conversation continued one of my associates expanded on the current state of affairs with the rank of battalion chief. "I've got a buddy who's a BC in City X, he says that most of their BCs have put letters in to demote themselves back to captains." Over the last few years a lot of the senior members of this department have retired. Many of the retiring captains are earning higher pensions than their BC counterparts. In some cases captains have pensions that are 30 percent higher. This did not go unrecognized by the current BCs or the crop of company officers considering taking the BC's test. This seems to be a common problem across the fire service. Many fire departments are finding it so difficult to get their company officers to take the chiefs test that they go outside the department to hire battalion chiefs. There are several factors responsible for knocking the luster off the BC's position.

In most fire departments BCs are sandwiched between the unionized rank and file and the department's senior staff, making them the proverbial middle child. Prior to promoting to the rank of BC my department ran a group of us through a BC's academy. Part of this training program included a four-hour session with the assistant chief of personnel and high-ranking member of our Human Resource division. I never knew what the HR

254

rep's exact function was but her favorite word was a firm "NO!" There were a dozen of us soon-to-be-promoted battalion chiefs in the audience. This is what we came away with (told to us in the form of stern lecture by the terse HR lady): "You are all promoting to the rank of middle manager. You will no longer be members of the bargaining unit. If you remain on a 56-hour shift schedule you will no longer earn 5 hours of comp time and 5 hours of time-and-a-half each pay period. You will no longer receive longevity pay and you absolutely cannot earn overtime. You will no longer have any say or input where you are assigned. Do you have any questions? The A/C of Personnel sat through her entire lecture with an ear-to-ear grin painted across his face. It felt like a swearing in ceremony for the Nazi Party.

The 12 BC candidates sitting before the two overlords just stared at one another for a few beats before erupting into chaos. Our organizational masters must have thought we would be as docile and bending as group of new fire recruits. Our collective group averaged more than 20 years on the job with 10 years as captains. Things settled down when the Lebanese Magpie took over as the spokesman for our group, "Thanks for the review. All of us knew that when we took the test. What the group of us wants to know is what are the benefits of being a chief? I've been trying to find out for the last 6 months and haven't got a straight answer. No one seems to understand the pay plan, performance-achievement markers or how much uncharged time we can take each quarter. Chief officers all receive a management and development fund but it remains a mystery about what we can spend that money on. I've asked payroll, personnel, and the city HR section what my starting salary will be and each one of them has given me a different answer. The group of us doesn't understand why this is so complicated." Within 2 minutes the HR shrew and personnel A/C were arguing over a set of answers that really didn't exist.

More than half of our group would take a pay cut the first year of our promotion. The answer to that one is pretty simple—captains who promote are very involved in the department. The majority of the soon-to-be-promoted BCs in my group were making anywhere from 20–40 percent over the base pay of a topped-out captain because of the special projects and other overtime endeavors we were involved in. Because the employer considers response chiefs as middle managers, they are no longer eligible for overtime. The problem with classifying a response chief's work hours in this fashion is they do not remotely resemble those of any middle manager. A response chief works a 56-hour work week—a middle manager works 40 hours a week. This is akin to pounding a round peg into a square hole.

A few years ago a group of us response chiefs were meeting with our boss. Two-thirds of us were there on our day off. We were wrapping things up when our boss (who worked a regular 40-hour week) announced we needed to have another meeting next week. He asked the group of us to pick a day. I suggested meeting next Saturday. He looked at us a little funny and innocently replied, "I don't work Saturdays." One of my off-duty counterparts said, "Yeah, I know what you mean. I don't work today but I'm here. Next Saturday works for me." Our boss smiled and told us how fortunate we were to have jobs and we should enjoy the extra meetings that came with it. Shift guys have never gotten any sympathy from staff guys when it comes to attending meetings on our days off.

When most middle managers take time off their position goes unfilled. This is one of the major reasons why middle managers are salaried and don't earn overtime. In most systems, when the middle manager works extra hours, they track it and take off a corresponding amount of time in the future. When a BC takes a shift off the position must be backfilled. City X strictly enforces the no OT rule for their chief officers. When a battalion chief position comes vacant company officers are hired back (on overtime) to work out-of-class. Now you have a situation where chiefs are ready and willing to work overtime but can't because of their middle manager classification. Instead, captains work in these positions because they are the only other available option. City management is loath to open the floodgates of overtime to response chiefs because of the possibility that other middle managers will begin demanding OT. There is a simple solution to this problem—put all 56-hour response chiefs back into the bargaining unit. This would free the municipality from the quagmire of having to create another layer in the middle manager ranks that really doesn't belong there. One of the standard arguments to keep response chiefs out of the union centers on the "conflict" they may have in disciplinary matters with members of the bargaining unit. If this were true the first rank that should be excluded out of the union would be the company officer. I supervise 4 battalion chiefs. Those 4 BCs supervise 45 captains. Those 45 captains supervise more than 250 firefighters and engineers. In the end company officers supervise more than 75 percent of a department's personnel.

A response chief's primary function is hazard zone management. This is a specialty that goes largely ignored in many systems. Most educational opportunities for operations chiefs are primarily administrative. It is absolute lunacy to send a brand new BC through a bachelors degree program in personnel management with a minor in budget analysis before you train them in incident management. Let's pretend for a moment that a fire department is like any other employer. The employer has an obligation and duty to provide a safe work environment for its members. Sending capable BCs to serve as the IC (or sector officer) at the scene of hazard zones is how the employer manages strategic level safety. Sending a BC who can't perform the command level requirements for a fast moving and unforgiving hazard zone is organizational Russian roulette. The department must prioritize the training requirements for new officers. First, train them in the things that have the highest risks and most severe consequences. A higher-level education is an asset for any officer but it must be in addition to being hazard zone competent if they plan on working in operations. This makes the rank of battalion chief (and shift commander) more akin to their rank and file brethren than the job classification of middle manager.

One of the common complaints you hear from chiefs is they have no control over their careers when they promote. If response chiefs had a set of assignment procedures that more closely resembled the rank and file it would go a long way in getting company officers to take the BC's promotional exam. It would also give chief officers more control over their career path. A few years ago the suggestion was made in my department that all staff chiefs be promoted to the rank of deputy chief. This would provide incentive for members to promote into staff positions. This would also lay the groundwork to have training tracks that actually match the different job qualifications for staff and operations chiefs.

Some consider it heresy to suggest that response chiefs should be rank file members of the IAFF. Wouldn't this be good for everyone? Some union officials don't want chiefs in the union because it blurs the line between "us" and "them". Some city officials and fire chiefs don't want chiefs in the union because it blurs the line between "us" and "them". Maybe including response chiefs into the union will create some unforeseen problems but I think the potential benefits far outweigh any downside. Response chiefs are in the most "operational" position in the fire department, having served in all three levels of the organization. Making a few simple changes to the BC's job classification will increase the number of quality candidates applying for the position. This leads to better bosses. Shouldn't this be a goal of both the union and the administration? After all, strong leadership is what separates good organizations from mediocre ones.

Chapter 31
Mothers

Today is Mother's Day, and it is my civic duty to recognize and pay homage to the people most responsible for propagating our species. I would like to thank my mom for birthing me and not throwing me off the Tallahatchie Bridge. She gave me sustenance and provided the biological blueprint for the way the neurons in my brain hook up. If the IAFC Human Resources Committee had taken this into consideration, they might not have been so hasty to declare me as an unwashed "Neanderthal," hellbent on destroying the fabric of the American Fire Service. Just like the other B-Shifters out there, I have a mommy who cared and nurtured me. I have fond childhood memories of my mother placing my younger brother in his "Little Driver" car seat. It was a hammock style seat that hung over the back of the front passenger seat, much in the same way a food tray hangs on a car door at a 1960s-era drive-in restaurant. The car seat had a small plastic steering wheel with its own little horn. The contraption gave toddlers the impression they could actually help mom or dad with the driving duties. Like many safety devices from the mid-20th century, it was fundamentally flawed. In the event of an accident, it would turn into a catapult, launching its occupant directly into the back of the driver's head, or thru the windshield. I miss the simple innocence of the 1960s.

After I became a firefighter, the "hearts and flowers" ideal I had of motherhood was shattered. My first engine company assignment had a first due that was equal parts pastoral and urban decay. It was one the few areas of town where one could find its residents swimming in canals that bordered busy roads. It wasn't unusual to drive down neighborhood streets littered with the decaying carcasses of large dead animals. The first due area boasted more broken down vehicles than ones that ran. During the three years I was assigned to this lovely patch of God's green earth our engine company delivered more babies than a Mexican midwife.

The first baby I had a hand in delivering was aboard a school bus. This wasn't a vehicle that transported students to and from school—it had been that in a former life. This school bus had no wheels and was parked in front of an abandoned church. We climbed aboard the bus/home and found a woman in her late teens, very pregnant and being tended to by her young man-mate. They were expecting their first child. Even at the tender age of 22 years old, I looked at the situation and asked myself, "What the hell are you two thinking?" This thought was quickly replaced when my training took over. Our captain requested an ambulance while the rest of us went about our job of taking vitals, boiling water and shaving the mother. 30 minutes later I watch as an impossibly large newborn emerged from a teenage girl's tiny orifice. I once heard a comedian describe this miracle of life as, "Watching a wet Saint Bernard squeeze through a cat door."

During the next few years our company delivered more than a baseball team worth of babies. In a small way we were helping motherhood. One day the miracle of life and the sanctity of motherhood bitched slapped our crew when we were dispatched to our monthly child-birthing call. Our first due area was littered with residences that had ½ addresses. These fractional street numbers were issued to housing units that had been built in the backyard of other houses. We arrived to the scene of a house that looked like a large two-story shed with a sitting porch. We navigated our way through a side yard to the back of the property where the ½ house awaited us. The residents of this tiny bungalow were way out in front of the green building movement. Their house was built of used tires, scrap metal, and railroad ties. Electricity was provided by a network of extension chords that had been wired into the electrical panel of the main house. The front door was open, so I knocked, tentatively stuck my head inside and announced, "Fire Department." A voice inside said, "Come on in, I'm back here."

The group of us took in the squalor as we entered what looked like a poorly kept fort for a tribe of 8-year-old boys. As we stood among piles of clothing, empty cereal boxes, a pair of crying children and a television with bad reception a woman yelled, "I thought I had to make a dirty, and the little bugger just popped out." The crew was exchanging glances of doom when a woman appeared in the hallway. She was wearing a pair of Buddy Holly eyeglasses, a T shirt and nothing else. She was walking bowlegged toward us because she had a crying, kicking baby dangling between her legs. The infant was swinging from the cord that ran up inside his mommy. The bottom half of mom and the entire newborn looked like a meat lasagna and a pitcher of tea had blown up on them. For a razor-sharp instant I prayed that the moon would collide with the earth and end it all.

Seven years later I sat in a child-birthing class with my lovely wife. We were a few months out from having our first child. The old battle-axe nurse instructor perked my interest when she picked up an empty apple juice bottle. She showed the big bottom to the group of us and said, "This is how big a new born baby is." She flipped the bottle around to show us the small end and said, "This is how big your birth canal is." The women in the group gasped. All of the soon-to-be new daddies wanted to beat the old nurse to death.

On our way home from the child-birthing seminar I was commenting on how inappropriate it was that several of the soon-to-be fathers planned on video taping the birth. My wife was of the same mind as me on this issue, but I lost her when I blurted out, "Can you imagine videotaping a rabid monkey chewing its way out of your life mate?" She accused me of not being excited about us making a baby. For the last part of the drive home we didn't say much. We were stopped at a red light when I looked up into the night sky and noticed the moon. It made me wonder what happened to the people who lived in the house made out of tires.

My firefighter career had left me jaded and over exposed to the biological side of life. This all turned around when I had my own tribe. The sea of estrogen that I swim in has washed me clean. I have come to learn that mothers are not only selfless and kind but they are the toughest people on the planet. A mother invented the cookie and they can produce their own milk. You can't ask any more from a person than that.

Chapter 32
Daddies Know Best

Last month I sent a B-Shifter thank you to mothers everywhere. Father's day rolled around a couple of weeks ago, and I meant to do the same for all the daddies out there. The staggering heat has sapped my paternal good intentions. It is so hot here that the river on the south side of Phoenix has been on fire for more than a week. It is so big and burned so long that they went ahead and gave it a name. During the last 10 days all I've been able to do is lie naked on the cold, concrete floor and quietly sob as my sweet bitches spray me with the garden hose while they text message their friends. I must make amends. Here's my belated nod to the daddies of our species.

I would like to take this opportunity to thank my father for the two greatest gifts he gave me. First and foremost dad, I am immensely appreciative that you didn't stuff me in a pillowcase full of rocks, tie off the end, and throw me into a canal. I've deserved it on more than one occasion. Secondly, thank you for not allowing mom to turn me over to the Catholic Church during my adolescence. I think that is enough about that.

Fathers are responsible for many of mankind's greatest achievements. While Bill Gates may have invented the VCR it was Al Gore who invented the Internet, which ushered in the golden age of porn. What do the two of them have in common? They're both fathers. Many of the world's great leaders have also been fathers. Seventeen of "Time "magazine's 20 greatest leaders and revolutionaries are men. I imagine most of them were fathers. Before any of you get the wrong idea that I'm somehow slighting women, consider that a high percentage of the people on "Time's" list fall into the category of despot (more than half of them, depending on how you feel about popes and presidents). This in no way implies that women don't make good leaders (or despots). In many ways, Margaret Thatcher was a modern-day Winston Churchill (both are on "Time's" list) but fathers tend to be the leaders of their family, tribes and cults.

Our thirst for top-flight leadership spawned the motivational poster industry. I am not a fan of the motivational poster. The last thing our world needs is another picture hanging in the workplace of an eagle in flight over the caption, "Leadership: Live daringly, boldly, fearlessly." If you added the phrase, "And while you're at it, Larry the Leader, why don't you blow a goat?" you'd have something. These insipid pieces of workplace propaganda have evolved to become profession specific. A series of these motivators have been aimed at the fire service—picture a weathered leather helmet above the awe inspiring words, "Command like a Champion." I miss the days when the only artwork hanging in fire stations were pictures of fire trucks, the occasional piece of military aircraft or posters of nude women riding tigers. "Command like a Champion" sounds like it came right out of the handbook on how to become a Scout Master. Can you imagine Chief Sitting Bull hanging that insipid drivel on the wall of his tepee? I think not.

1988 marks the year when my virgin eyes were

assaulted by motivational wall art in a fire station. I was a new captain and had just been released from my first temporary assignment. One of my contemporaries, the freshly promoted Captain Serge, and I were reversed into a multi-company station after both of the previous captains had been organizationally evicted. This station was an organizational wart. Their final dildonic act was an ethnic skit performed for an ethnic firefighter. This practical joke's punchline led to suspensions for several members of the station along with the mandatory expulsion of both the station's officers. Seniority was the reason the department's two most junior captains ended up having to deal with the remnants of the organization's version of the "Jerry Springer Show." We had no seniority, so we went. The next three months passed very slowly. During that time I never felt bold, fearless or daring —only like a baby sitter sent to watch a bunch of afflicted monkeys.

After my temporary assignment ran its course, I went roving again. My first shift of freedom found me filling in at multi-company station just west of monkey-central (my former temporary assignment). I put my gear on the engine, checked out the rig and then headed into the station. I was mobbed as soon as I got into the kitchen. Everyone wanted to talk about the juicy details of my time spent with the monkeys. I shared the nickel version and then went on with my morning. I hauled my bedding into the captain's office/dorm where I met the ladder captain. He pointed me to my bed. As I threw my stuff on my rack he kicked the door closed and asked me to sit down. As I sat down I noticed the wall-length bulletin board behind his bed was plastered with motivational propaganda.

This was during the late 1980s, when the motivational poster industry was still in its infancy. Much of the tacked up art was homemade and had the look of ransom notes. The lunatic officer I was currently locked in a small room with had found a bunch of pictures that inspired him and glued words on them that he had cut out of magazines. The only non-homemade item was a poster of a pair of firefighters silhouetted against a fire. The caption read, "Adversity introduces a man to himself." The rest of the wall was covered with pictures of sunsets, sporting events, bears, mountains, space ships and soldiers. There wasn't a single tit to be found anywhere. It was an altar to self-help through arts and crafts. Captain Motivation wasn't widely known for his inspirational art fetish. The event that most defined his career was having vented the wrong roof. This happened several years earlier, when Captain Motivation ran the ladder company out of the monkey house. This wasn't a case of tactically missing the correct hole placement on a building with a series of adjoining roofs. Captain Motivation and his monkey ladder crew had cut a hole in the roof of an exposure that was more than 100 feet away from the fire occupancy. There wasn't a poster on his wall that addressed venting a perfectly fine Chinese food buffet across the parking lot from a burning hardware store.

As I sat down and faced my new roommate I didn't know what to say, so I didn't say anything. He looked me up and down like he was about to guess my weight before he said, "I want the real story about what went on at Station #? (aka monkey central).

I shrugged and replied, "It was a long three months."

"Were you and Captain Serge required to stay there until you ran everyone out of the station?" he asked.

"No, it just kind of happened that way. The group of them refused to behave like humans. After a while, Serge and I just got sick of their shit and documented their idiocy."

"I don't think the department did anyone any favors with the way they handled that mess," he said. "The two most junior captains shouldn't have been sent there."

"Both of us voluntarily went through the process to become captains," I stated. "No one else wanted to go there. We were the two most junior captains. We went. That's the way it works."

Captain Motivation looked over the top his glasses and made a bitter granny face when he condescendingly replied, "Please don't take this personal, but neither one of you guys have the experience to deal with those types of personnel issues."

"I understand the point you're trying to make," I said. "But when you consider that the monkey-central group had been feeding off one another's dysfunction for almost a decade, your theory falls apart. Hell, you know what they were all about. You were captain at that station for five years and couldn't do anything about it. They ended up running you out." I felt our meeting was in its wind-down phase when Captain Motivation said, "At least you had the advantage of working there with a like-minded captain and had the full backing of the department. I was up against the whole station and didn't have any organizational support."

I didn't plan on saying anything, but the little cocksucker who lives in my temporal cortex pulled the string that caused me to blurt out, "I guess you could say that adversity introduces a man to himself."

Chapter 33
Accountability Pimps

I quit watching the news several years ago because the whole experience agitated me. Most of the anchors exuded an air, as if their essential body fluids had been replaced with a combination of corn syrup and the blood of Christ. The weather guy was just too excited about reporting the weather ("It's so hot it's made me queer!"), and the sports guy yelled too much and looked like he beat his wife. A few nights ago I made the mistake of walking by the TV when my wife was watching the nightly news. The lead story was the effect sky-rocketing gas prices had on America's driving habits. We are driving less. In May of this year we drove almost 10 billion fewer miles than we did in May of 2007. The tone of the story left me feeling as if there was something wrong with all of us. Halfway through the story they showed footage of last year's bridge collapse in Minneapolis that killed six people. The narration cautioned if we continue to buy less gas, the government will take in fewer tax revenues and more bridges will fall down. Before the story ended I ranted at my sweet love pie, "This is total bullshit. On every other channel they're berating us for using oil. We should grind all these cocksuckers into pulp and feed them to our dogs." I believe the media has evolved into a vindictive ex-spouse. One that perpetually nags that we can't do anything right.

None of this hooks up to reality. Nothing is discussed or debated anymore. The right wing of the green party hails Al Gore as a savior and then attacks the rest of us for having our groceries bagged and taking showers. It seems like the attitude has become, "You're in favor of flying around the world 100 times a year to tell us to turn the lights off and to quit using carbon-based fuels." The new obsession is global warming. The environmental zealots scream, "We are killing our mother!" This hysteria is more contagious than any virus. None of us wants to anger the popular kids so we just nod and go along. It is human nature and is as old as the monkey.

The herd mentality is also the basis for commodities and free markets along with every other Ponzi scheme ever invented. The collapse of the housing market, global warming and the insane price of oil have a lot in common. The average price of a house in Phoenix Arizona used to be around $150,000. Then one day the group decided it was a seller's market and the price went to $155,000. Pretty soon, the human home-buying herd figured there were unimaginable riches to be made in the real-estate market and buyers were having bidding wars in the front yard of a formerly $150,000 home. Before you knew it, human fear and greed artificially inflated the value of $150,000 beyond the $200,000 mark. This went on until some expert declared that the housing market was overpriced. The media over-reported it, the herd panicked and the housing bubble burst. The same thing happens on a regular basis with gold, oil, grain and every other thing that is bought and sold.

The theory of global warming is no different. The temperature has been steadily rising since the last ice age.

Scientists now believe the increase is due to the amount of CO_2 we are pumping into the atmosphere. This frenzy all began when a few scientists said, "It's getting hotter." After a while some more scientists joined in, and before you knew it, the scientists were overtaken with global-warming hysteria. Because scientists are the smartest people in our society, the media and pop icons banded together with them to tell us we are killing our mother earth and our children will boil in the excrement created by their parents. Before you rush out to buy a fistful of carbon credits, keep in mind that less than 400 years ago Vatican scientists threw Galileo in jail for suggesting that earth wasn't the center of the universe. This was during a more innocent time, a time when religion, science and government could be found under one roof.

The government has always used science and statistics to their benefit. A few years ago the federal government financed anti-drug commercials warning the viewing public that buying marijuana supports terrorism. Let's review: Most of the marijuana that makes its way into the United States comes from Mexico, Columbia, Jamaica and Canada. Our government has never listed any of these herb-producing nations as a terrorist threat to our country. In fact, more than half of the marijuana used by the US stoner population is domestically grown. If you marry the propaganda of the government's anti-drug campaign with the facts, it would appear that over half the money Americans spend on wacky-weed is funneled directly into American terrorist organizations. The criminalization of marijuana becomes truly obscene when you consider that cilantro is widely available in every grocery store across this great nation of ours. It is a foul weed that pollutes the food we eat and it is 100% legal to import across our border. When will the madness end?

While I'm on the subject of illicit drugs and the current church-controlled dialogue that surrounds them, let's take a minute to study the dynamic between opium and an actual government who supported terrorism—the Taliban. Before the Taliban controlled Afghanistan, that country accounted for 75 percent of the world's opium production. The Taliban tends to be extreme fundamentalists and felt opium production didn't fit into their religious dogma, so they outlawed poppy cultivation after they seized power. Within a few months, they eradicated more than 96 percent of the poppy fields in the country. They also outlawed flying kites, dancing, music, TV, shaving and the blowjob (even between a man and a woman). I guess one of the lessons we can take away from this is when you get good at eradicating stuff it can be pretty tough to stop.

All of these wackos preach the common mantra of "accountability." They expect total obedience from the masses. You don't have to look any further than your backyard to find the over zealous personal conduct police. They exist in every organization and are a fact of life. Most of the accountability fanatics I've worked with had the moral code of a third-world pimp. It's fascinating to watch the evolution of these individuals as they promote through the ranks. Their expectation of others seems to rise in conjunction with their rank. Another symptom that goes along with this affliction is a loss of memory. This ends with a funky form of delusional absolution. Having a high-ranking chief teach a group of underlings an ethics class is a pretty trendy thing to do. The class takes on a whole different vibe when you consider that your ethics teacher nailed his secretary doggy-style, on top of his desk, during his lunch hour (technically it wasn't on company time). It's every bit as ludicrous as a conservative U.S. senator who publicly denounces same-sex marriages getting caught soliciting gay sex in an airport bathroom.

Another facet of the accountability disorder is the way it can distort your concept of competence. The only reason a fire department exists is to go on calls. Our core competencies must be designed around performing the work. As you promote, your job shifts to somehow support, manage or lead the work. Issuing the battalion chiefs color swatches to go out and check the coloration of the workforce's uniform T shirts falls on the medication-worthy side of compulsive. The struggle is to differentiate between what's important and what isn't. For bosses to do this effectively, they must understand the full scope of the work and what is required in the way of training, tools, equipment and support for the workforce to go out and do the work. This isn't as easy as it sounds. When the bosses don't understand the work, or even worse—they minimize the work because it's "beneath" them, they tend to become hypnotized by the concept of obedience. Now the leaders of the organization have their priorities completely fucked up. The No. 1 organizational directive becomes following the minutiae of rules and regulations: Can we wear our ball caps forward or backward? How many piercings can a senior engineer have in his lip? What's an appropriate amount of tattooing a member can display below the sleeve of their shirt? And the ever popular: Can a member blend two shifts of sick leave with a Kelly day, three shifts of vacation and two AWRs and take three weeks off? Personal conduct and grooming standards are oftentimes the bastion of the incompetent boss, because managing the real work requires authentic competence. We aren't a fashion house or a boarding school; we're a fire department.

All the bullshit will drive you batty if you let it. Consider that over the eons our mother earth has been bathed in both ice and fire and the concept of global warming is put into a more complete perspective. Politics, law making, and the news are vexations best left to those who can't do anything else.

Chapter 34
Assault & Battery with a Christmas Tree

The holidays are upon us. Ten years ago I was asked by the owner of a now defunct, used fire apparatus monthly periodical to write an article about my first fire. I shared the true tale of my brother and I disposing of our family Christmas tree. I believe less than 50 people read the original airing of the story. The Christmas tree story grew a wider audience when it appeared in the pages of a fire-service trade journal a couple of years later. Christmas season is upon us once more, and the Christmas tree story has run its course. I am at the point in my career where I've also grown weary of fables featuring firefighter Santas who cure crippled children simply by touching them with the antenna of their portable radio. The following event happened on the night my oldest daughter was conceived, and it also involves a Christmas tree.

Twenty-one years ago a group of us decided to have a holiday festival. We rented out a party hall, two floors of a downtown hotel, hired a caterer and procured enough liquor to kill a small European nation. After settling on a date we did something that was unusually intelligent and mature for a herd of B-Shifters—we arranged transportation services from the party location to the hotel. The party went off without a hitch. Over one hundred B-Shifters and their wives and girlfriends showed up and ate, drank and danced. Everyone behaved reasonably well and we got our deposit back from the party hall. Around midnight an assortment of limos and taxi cabs arrived to usher the group of us to our hotel where the holiday festivities moved into its second stage.

After arriving to the hotel the party resumed in one of the hotel's conference rooms. A bar was set up in the corner of the room, and within the span of a few minutes the room went from comfortable to packed. A small group of us were crammed in a corner, shouting at one another to be heard over Lynyrd Skynyrd wailing about "that smell." We were less than 30 minutes into the hotel party when the initial disturbance happened. One of the women in our conversation circle fell to the floor. This would be her first (and last) date with a B-Shifter. She was very polite and attractive. As I remember, she was somehow related to an A-Shifter from another department. It was one of those "wife's best friend/ husband's best friend" arrangements that always end in disaster. The poor thing had been pulled to the ground by a young firefighter who was using her as a ladder in a vain and pointless effort to stand.

Otis El Borracho was a firefighter with one whopping year of seniority, yet his reputation was well known across a fire department that protected a 500-square-mile city. The "Otis" came from the lovable Otis Campbell character from The Andy Griffith Show. Andy, Barney, Aunt Bea, Opie and Otis the Drunk—it was required lunchtime viewing in every Phoenix Fire Station. El Barracho quite literally means "the drunk" in Otis's native tongue. Several of us pulled Otis off the woman, diagnosed him with exceptional drunkenness and carried him to his room. We dumped Otis's 250-pound bulk into his bed, turned him on his side in an effort to avoid a Janis Joplin-style death and headed back to the party.

We were having a fabulous time when an agitated desk clerk barged into our event, demanding that someone come down to the lobby, "To deal with a psychotic naked fireman!" We should have known better. When we got down to the lobby, we found Otis sprawled across a bell boy's luggage cart. He was shirtless, had his pants around his ankles and was covered in potting soil. He had used the luggage cart as a battering ram on several large potted plants in the lobby. We had a brief conference with the night manager and determined we would not be getting back our security deposit. Management's only concession to us was if we got rid of Otis the rest of us could stay. As luck would have it the only sober party-goer had his vehicle at the hotel. We loaded Otis up and hit the road without much of a plan.

The three of us were in a pickup truck. Otis rode in the middle and was flopping back and forth in an alcohol-fueled coma. My partner and I decided that we should drop Otis off at LARC and be done with him. LARC stands for Local Alcoholic Rehabilitation Center. When we responded to calls involving drunks sleeping on the sidewalk or wandering around in traffic, we'd call LARC. They'd show up in a wino paddy wagon and haul these artists back to the LARC facility, where they could sleep it off, wake up to a delicious baloney sandwich and a one-on-one with a burned out counselor. Our only problem was we had no idea where the LARC facility was located – they always came to us. It was about this time that Otis woke up long enough to projectile vomit all over the windshield.

In less than a minute I had my window down and was holding Otis by his feet. He was hanging out the window making death noises. It was as unpleasant as it was unsafe, but the retched bastard wasn't getting to ride inside anymore. Otis worked at a fire station located a half mile from our current position and our best course of action would be to drop him off there and be done with him. Our plan was to lean him against the door, ring the night buzzer and take off. As we watched chunks of vomit fall off the windshield we amended the plan to include a quick clean up of the cab of the truck. The situation went from bad to worse when we arrived to Otis's fire station.

Station 99 was a 1950's era, red brick fire station. It had a shallow front yard and both the station's bay doors were open. The apparatus bays were empty, telling the world that the engine and ambulance were out. My buddy had been growing increasingly pissed off about his vomit-strewn dashboard. This became apparent when he slid his truck to a stop in Station 99's front yard, clipped the flagpole and came within a few inches of hitting the front of the station. I let go of Otis's feet, opened my door, stepped out of the truck and noticed the station's flag was still up. Otis was crawling across the ground on his back, hallucinating that I was his wife. He was shrieking at me that I was a no-good "punta" and was responsible for destroying his life. I grabbed him by his vomit-drenched shirt and hauled him to his feet and lead him toward the dark, empty bay. As I reached the front of the station I grabbed the garden hose and turned on the spigot so I could rinse Otis off prior to depositing him into the station.

The stream of frigid water panicked Otis. He blindly grabbed at the source of the assault. I pulled the hose back, pushed him away and directed the stream in his face in one fluid, ninja-like motion. We were now in the middle of the apparatus bay dancing in circles. Otis was doing his best to drown and he was trying to take me with him. My partner was shouting for me to give him the hose when the front of the station lit up and a voice yelled, "Freeze!" As I turned to look Otis kicked at me, lost his balance and fell on the wet floor. The voice from the lights shouted, "Drop the hose now!" As my eyes adjusted to the light I saw a cop perched in a combat stance with her gun pointed at me. I shouted back, "I'm an EMT, and I'm saving this man's life." I felt confident she wasn't going to shoot me because Engine 99 was in the process of pulling into the bay, screwing up her line of fire.

The captain came off the truck and calmly asked, "Why are you watering Otis at 1 in the morning in the middle of our apparatus bay?" As my partner walked up he answered the captain's question, "We're dumping him with you guys because he was attacking the women at our party. After he tackled my date, he busted up the lobby and was generally making a spectacle of himself. On the way here he puked all over the inside of my truck. If you don't want him maybe the cops will do all of us a favor and throw his amateur, drunk ass in jail." I turned back and saw half a dozen police cruisers parked helter-skelter in front of the station. The cops had assumed a relaxed posture, indicating the three of us didn't pose a serious threat to the C-Shifters of Station 99. A female sergeant wearing black gloves stepped forward while holstering her service revolver. She looked at Otis and asked, "How much liquor does it take to do that to such a big, strong firefighter?"

The cops lost interest after determining they wouldn't get to shoot any of us. They also took a pass on transporting a large, soaking wet B-Shifter who was prone to vomiting fits. The C-Shifters agreed that Otis could spend the night on the dayroom sofa if we cleaned him up. I gave Otis one final rinse and put him night-night while my counterpart cleaned his truck. Within a few minutes Otis was snoring on the sofa. The blinking lights on the station Christmas tree threw a red and green pallor over the ambo sheets wrapping his body. He looked like a Mexican Caesar in a Fellini film.

When the two of us returned from our hour-long errand the party had spread throughout the hotel like a virus. The front desk was deserted. I ran into my wife on the way to the conference room. We headed back to our room down a long hallway where the atmosphere was filled with the dying remnants of dry chemical powder. It looked like the fog had rolled in. Laughing, screaming, arguing and sobbing from throughout the old building all merged to give the hotel a haunted feel. We made it back to our room without incident. Nine months later we were blessed with the arrival of our first she-cub.

A week after our party I ran into Otis on an EMS call. He was covered with a collage of small cuts and abrasions. When I asked him how he got them he grew somber and reflective. Paramedic Bob on the other hand grew animated. He slapped Otis on the back and said, "You don't remember any of this, and I will never tire of telling the story. After you put Otis to bed, we made it back to the station. I was working an overtime shift on the rescue. The C-Shifters filled us in on how Otis ended up looking like a dying mummy on the station couch. A few hours later we all got back from a call and the ambo guys starting screwing with Otis. They turned on all the lights in the dayroom and screamed that Otis was missing a call. Otis came off the couch running and tripped because his legs were wrapped up in the sheets. Now he had six C-Shifters laughing their asses off over him. It was obvious that he was still fucked up beyond imagination as he thrashed to get out of the sheets. When he finally tore his naked body free from the sheets he fell into the station Christmas tree. This had the C-Shifters hyperventilating with laughter. Otis staggered to his feet, grabbed the Christmas tree by its trunk and began beating the C-Shift ambo crew with it. Ornaments were exploding and Otis's naked body was covered with garland, tinsel and blood. The C-Shifters were running around and screaming like little girls with their hair on fire. It was like something straight out of the bible."

Merry Christmas to all, and to all a good night.

Chapter 35
A New Emanicpation Proclomation

My former employer used to have an employee suggestion program. For all I know they may still have this program, but I quit it using after I got no response from the first two suggestions I sent in. I really can't blame them for the first one. I thought it would be a swell idea if the fire department had an employee of the month. My suggestion included three awards for this honor. First, they would get two extra weeks of paid vacation, second, they wouldn't have to drop for kitty for the month, and third—a full compartment door on every engine and ladder company in the Phoenix Fire Department fleet would have a picture of our employee of the month. I still get a giggle imagining firefighters armed with Sharpies drawing blacked out teeth, arrows through the head and giant penises hovering toward the employee of the month's mouth. It would have been a hoot.

My second suggestion should not have been so hastily thrown in the trash. The beauty of this suggestion is its timelessness. Let me begin by giving you some background. I spent the majority of my career working within a mile or so of Van Buren Street. Other than the mile or two that bisects downtown, Van Buren can best be described as not meeting its full potential. Gangs, illicit drugs, homelessness and prostitutes abounded from one end of Van Buren to the other. Every city of a certain size has their own Van Buren Street. Phoenix is big enough that it has a couple of them (despite what the people from the Chamber of Commerce and Department of Tourism tell you).

My suggestion was to create an enterprise zone for a half mile in each direction from Van Buren's center line. This mile-wide corridor would carry special zoning that allowed the building of casinos, brothels and hash bars. I reasoned that all of these activities already existed in abundance along this stretch of the city. At the time of my suggestion, the state was negotiating with the Indian Tribes over building casinos on their reservations. All of the cities in the Phoenix metro area were also trying to outdo one another with tax incentives and other big giveaways to lure business into their tax boundaries. I argued that the residents of Phoenix would be the largest customer base at the soon-to-be-built Indian casinos. The untold profits would go to the tribes while any taxes and fees on these establishments would go directly to the state. I found this notion crazy. The city leaders would be shitting baby cacti if all the local car dealers decided to pack up and move their dealerships to the Indian reservations but the whole sorry group of them simply yawned at a state industry that has produced billions of dollars of profits in 15 short years.

Opponents of local gaming use the argument that they don't want their beautiful city turning into Las Vegas. That's pretty simple to get around—don't build Las Vegas-style casinos. Write the building code so these establishments have to look like English castles, French gothic churches or any other architecture that will make your community even more beautiful. The people who build and operate casinos could give shit what they look like so long as they can pack the inside full of games of chance and a never-ending line of people willing to bet the farm on the next turn of a card. The "It degrades our community" argument completely loses traction when you factor in the Van Buren Street I lived and worked on every third day for 20+ years.

My proposed enterprise zone contains the state hospital for the criminally insane, the regional drunk tank, crack houses, low-rent titty bars that were nothing more than fronts for prostitution and the heroin trade, homeless shelters, an abundance of halfway houses, used car lots, grocery stores that sell whole animal heads, after-hour social clubs, recycling and wrecking yards, gay bath houses and, for a short period of time the, Arizona Republican Party headquarters. It is a place that your mother, priest and airport skycaps warn you not to visit. My three-part plan would legalize gambling; decriminalize prostitution and all forms of cannabis. These activities would be restricted to the Van Buren enterprise zone. I had included all drugs in my original plan because in the course of a busy shift we ran on heroin overdoses, tweekers, drunks and crackheads. We once assisted a man who had consumed peyote and washed it down with a bottle of Tidee-Bowl. He told us, through blue lips, that he thought it was grape cough syrup. I reasoned that despite being illegal, the natives had quick and easy access to their drug of choice. Why not make them legal? Control them, tax them and take them (and their profits) away from the criminals. If you remove the criminals from the equation it can't help but to drastically cut crime in the area. The city could use the newfound revenues to build parks, send the mayor and council on high-roller junkets, spend obscene amounts of money on controversial public art, and pay health insurance premiums. We all win. I realized that including all drugs was probably too big an idea for your average politician to wrap their head around so I amended my plan to only include all cannabis products. This was a good starting point and provided a beachhead (a gateway beachhead) for future legislation to include all drugs.

I never heard back from the employee suggestion people. Let's fast forward to a scant few months ago. Van Buren Street is still a destination for the broken who live on the ragged edge of society, and the word on the street was the city was looking at a $270 million budget shortfall. After the budget experts put their pencils to it, they came up with a new deficit figure of $190 million. Somehow $80 million was magically milked back into the ledger. I quit paying attention to any budget propaganda when I figured out that the Budget and Reporting Office had access to a bunny that lays $80 million dollar eggs. Meanwhile our governor and state legislature are a month overdue in coming up with a state budget. They are only $3.5 billion in the hole. The federal government's deficit is beyond comprehension. The last I heard is every American household will have to pony up $500,000 to settle the debt. None of this seems to matter as the feds are wearing out the presses that mint money. If Obama were a true visionary he

290

would send every American household a million dollars and let the chips fall where they may. It isn't any crazier than their current course of action.

It wasn't that long ago that our last president urged us to all go shopping. Hell, he even sent us all $1,000 in the mail to get us started. The shrieking noise that fills the air today is Susie Orman scolding us to watch our spending and maintain cash reserves that can carry us for a year. That's just what we need. Unemployment is through the roof, people are losing their homes, parents can't send their kids to college and some arrogant bitch who pulls down millions of dollars a year lectures the unwashed consumer. Who the fuck is this woman? This is no different than shouting at a man being eaten by a shark, "You should learn to karate swim!"

It is becoming painfully obvious that our financial leaders are nothing more than a bunch of well-dressed used car salesmen. I'd never thought I'd live to see the day when I'd be a better financial risk than Citi-Bank. Insane monkeys, wearing striped slacks and plaid jackets couldn't do any worse. Bernie Madoff is a pariah because he swindled investors out of a reported $51 billion. The real question is where is all that money? Saddam Hussein and his two crazed sons (God rest their egomaniacal souls) couldn't spend that much money. Madoff gets ushered off to jail while the executives responsible for the loss of trillions of dollars fly in private jets while they finagle bonuses that would make King Solomon blush. Now the filthy rich who own and operate the health insurance industry have scaled the tower to warn us about the impending federal takeover of the health-insurance industry. This is a group who takes 30 cents on every dollar spent on health insurance. Dealing with private health insurance companies feels a lot like dealing with the IRS so the federal takeover should be pretty seamless. I'll be more sympathetic with their vapid arguments about socialized medicine as soon as the health-insurance profiteers take a step back and remove their dicks from the consumer's ass. Here's a novel idea: Get a room with Susie Orman and spend the rest of eternity counting your money.

The whole financial system has evolved into nothing more than smoke and mirrors. Our country's early financiers swindled the Indians out of what would later become Manhattan for a pocket full of beads. Now it has come back to bite us in the ass. The only recent legislation that makes a lick of sense is a bill that's making its way through the California Legislature. Some demented politician wants to legalize marijuana in the state of California and tax it. Never mind that the state spends millions of dollars a year imprisoning people whose only crime is the possession or sale of reefer. Never mind that the state spends millions of dollars a year on law enforcement's pointless attempt to eradicate the state's top cash crop. Never mind that marijuana is a much safer substance than tobacco and alcohol ("You don't want none of this, Dewey!"). Never mind that marijuana has slowed and actually reversed some forms of cancer. Never mind that the pharmaceutical, liquor and cigarette companies spend millions of dollars every year to keep marijuana illegal. They don't want that kind of competition. I think the real issue here is America.

The one consistent ideal we use to describe our country is freedom. It's implied in our pledge of allegiance and every song ever written about our land. What true freedom boils down to is this (and it is not my intention to go Southern on you, but this is the best definition): If you don't fuck with me, I won't fuck with you. I no more endorse the smoking of dope than I do drinking hard liquor, watching Nancy Grace, stockpiling automatic weapons, fouling good food with cilantro or two men blowing one another. As long as you don't get into my business, I won't get in your business. That is the foundation of a civil society and the cornerstone of true freedom.

The magpies who constantly nag for moral purity sound more and more like every other group of extreme religious fundamentalists, the only difference is the language they shriek in. We are not a Christian nation —we are a free nation. Free to pray to the god of our choosing, free to pursue the career of our choosing and free to live our lives. It wasn't that long ago that a black person wasn't free to drink out of the "wrong" water fountain, today gay people are only free to get married in a few states and none of us are free to pay for pussy and smoke a big fat daddy. If the government would get out of adults' personal business, this would lead to a more laid-back and financially secure country. We need a emancipation proclamation for the 21st century.

The Puritans warn that if we legalize prostitution and drugs, what's next? Here's a novel idea: taxing the church.

That's what's next. If the Christian right can take over the Republican Party then they can go ahead and pay for the right to own part of the electorate. Pick up any paper and you can read lurid tales of our elected leaders paying stupid money for hookers, sexually harassing congressional pages, having love children outside of marriage and soliciting gay sex in public bathrooms. I am sick and tired of these sinners evoking god in their post "I got caught" press conferences. Just come clean. It would be refreshing to hear just one of them say, "I was in a situation where I took advantage of the pussy that was offered to me. If there aren't any more questions, I'll be getting back to work." These are the same hypocritical, two-tongued hunchbacks that have handed the country over to the highest bidder. They are shortsighted morons. Legalize it, tax it and watch the money roll in. We will pay the debt off in our lifetime by simply controlling activities that most of congress is partaking in right now.

The last three American presidents have used recreational drugs, and it appears to have adversely affected only one of them. The problem with making things illegal is it hands entire industries and huge parts of the economy over to criminals. If you want an example of criminal run businesses take a trip to the U.S./Mexican border and count the heads on the side of the road. We've got to quit maintaining a set of unenforceable laws. No one pays attention to them and they diminish the important laws that we all need to follow. This is the same concept as a fire department having a small set of really good SOPs that everyone follows versus maintaining a extra large set of shitty and out of date SOPs that no one follows. Effective and well-crafted SOPs (or laws) streamline and enhance operations. My former department's SOPs were taken to the next level when we opened a Command Training Center where we could actually test our SOPs against the most realistic training we'd ever done. This project was so successful that it was shut down the day our former "operational" fire chief retired and was replaced by a set of political administrators dressed up like firemen.

The days where one could exact political revenge for fun are rapidly being replaced by the need to amputate people and programs because of the ever-shrinking budget. The current budget crisis is making it tougher to keep the machine moving. This summer Phoenix shuttered all of its public swimming pools on the hottest day of the year. The reason—we're out of money. It had to be tough on the mayor. He touts himself as a family values kind of guy on the news and in all of his election propaganda. Political myth and reality rarely mesh. Legend has it that his real family refused to pose with him for the picture he wanted to use in all of his election advertisements. This didn't slow him down even a little; one of our high-ranking union officials loaned the mayor his family for picture day. The political gene pool is very shallow, so be extra careful if you decide to dive in.

I hear our nation's most rational leaders proclaim that these troubled times will require intelligence, patience and ingenuity. I believe it is time to quit shitting ourselves. If we plan on keeping all of our eggs in a free enterprise economy it is time to make some changes. Many cities are out of options to deal with their current financial crisis. Last year the entire country of Iceland went bankrupt. Hell, it's so bad there that McDonald's pulled out.

Our country throws away more than $6 billion a year to incarcerate non-violent drug offenders. Who among us would write a $25,000 check every year, for three to 20 years, to cover the prison costs associated with locking up the neighbor down the street for cultivating a marijuana plant in his backyard? Based on most economic indicators, any competent accountant (who wasn't employed by Enron's former firm of Arthur Andersen) would declare that we are broke. We can't afford to keep living the illusion that we're a group of Salem witch hunters on a self-proclaimed mission from God. Wake up and embrace the new industry, and before you know it, we can revitalize all of our country's Van Burens into cosmopolitan Mecca's that will make religious extremists green with envy.

CPSIA information can be obtained at www.ICGtesting.com
Printed in the USA
LVOW130351050113

314487LV00004BC/5/P